AF508680

Current Research in Thin Film Deposition

Current Research in Thin Film Deposition

Applications, Theory, Processing, and Characterisation

Editor

Ross Birney

MDPI • Basel • Beijing • Wuhan • Barcelona • Belgrade • Manchester • Tokyo • Cluj • Tianjin

Editor
Ross Birney
University of the West of
Scotland
UK

Editorial Office
MDPI
St. Alban-Anlage 66
4052 Basel, Switzerland

This is a reprint of articles from the Special Issue published online in the open access journal *Coatings* (ISSN 2079-6412) (available at: https://www.mdpi.com/journal/coatings/special_issues/ film_depos_appl_theory_process_charact).

For citation purposes, cite each article independently as indicated on the article page online and as indicated below:

LastName, A.A.; LastName, B.B.; LastName, C.C. Article Title. *Journal Name* **Year**, *Volume Number*, Page Range.

ISBN 978-3-0365-0512-1 (Hbk)
ISBN 978-3-0365-0513-8 (PDF)

Cover image courtesy of Ross Birney.

Contents

About the Editor

Ross Birney is a Lecturer at the University of the West of Scotland and received a PhD in Physics from the University Of The West Of Scotland. Dr Birney has over 10 years' experience across a broad range of thin-film and advanced materials research spanning both industry and academia, in the UK and overseas. His research interests include optical coatings, gravitational wave detection, protective/anti-corrosive coatings, amorphous carbon, and ion beam technology. Link to PURE: https://research-portal.uws.ac.uk/en/persons/ross-birney LinkedIn: www.linkedin.com/in/ross-birney ORCID: https://orcid.org/0000-0002-4766-0757Google Scholar: https://scholar.google.com/citations?user=kQN55PoAAAAJ&hl=en.

 *coatings*

Editorial

Special Issue: Current Research in Thin Film Deposition: Applications, Theory, Processing, and Characterisation

Ross Birney

Institute of Thin Films, Surfaces and Imaging, Department of Physics, University of the West of Scotland, Paisley, Scotland PA1 2BE, UK; ross.birney@uws.ac.uk

Received: 11 December 2020; Accepted: 14 December 2020; Published: 16 December 2020

Today, thin films are near-ubiquitous and are utilised in a very wide range of industrially and scientifically important areas. These include familiar everyday instances such as antireflective coatings on ophthalmic lenses, smartphone optics, photovoltaics, decorative and tool coatings. There exist also a range of somewhat more exotic applications, such as astronomical instrumentation (e.g., ultra-low loss dielectric mirrors and beam splitters in gravitational wave detectors, such as Laser Interferometer Gravitational-Wave Observatory (LIGO)), gas sensing, medical devices and implants, and accelerator coatings (e.g., coatings for the Large Hadron Collider (LHC) and Compact Linear Collider (CLIC) experiments at European Organization for Nuclear Research (CERN)).

The aim of this Special Issue is to provide a platform for researchers working in any area within this highly diverse field to share and exchange their latest research findings.

Thin films are everywhere in the modern world, with many of the technologies we depend upon in daily life being, in turn, dependent upon thin film technology. These may range in dimension from an atomic or molecular monolayer—perhaps only a few ångströms thick—to either mono- or multilayer coatings with a thickness of several microns.

Such materials may have a huge range of extremely useful properties; they may be, for example, anti-reflective, impervious to oxygen and/or other gases, optically transparent yet electrically conductive, catalytic, and self-cleaning. Everyday examples featuring thin film technology include, but are not limited to, mobile phones, touch screens, laptops, and tablets [1,2].

Other important applications of thin films include bandpass filters as used in gas analysis [3], mirrors used in astronomy [4–6], protective (e.g., biomedical, anticorrosive, and antimicrobial) coatings [7], architectural glass coatings (e.g., to reflect heat while transmitting visible light) [8], photovoltaic electricity generation [9,10], and a great many others.

There are also a great many routes to deposition of thin film materials, including electron beam evaporation [11,12], ion beam sputtering [4,6,13,14], chemical vapour deposition (CVD) [7,15–18], magnetron sputtering [19–21] and atomic layer deposition (ALD) [22,23].

As a result, thin film deposition continues to be a very active area of research and development.

This Special Issue contains 11 original research articles.

In their article, Hseih et al. [24] report on the effect of hydrogen dilution ratio ($R = H_2/SiH_4$) on the structure and optical performance of nano-crystalline hydrogenated silicon (nc-Si:H) thin films, deposited by plasma-enhanced chemical vapour deposition (PECVD). They describe the process by which the transition from the amorphous to nanocrystalline state can be controlled by careful adjustment of R and suggest an optimal range of $R = 30$–40 to favour nc-Si:H growth. This is likely to be of great interest within the fields of photovoltaic energy generation and optoelectronics.

Schwinger et al. [25] discuss a novel, non-vacuum, liquid phase method of depositing thin films of aluminium on spherical glass substrates, comparing the results to those obtained by a more conventional physical vapour deposition (PVD) process for reference. It is reported that reflectance of solid micro glass spheres can be increased by around 30%, a comparable result to that achieved using PVD.

This could have applications in, for example, road stripe paints and in mid-infrared reflective interior architectural paints, which can potentially reduce energy usage for the heating of buildings. Li et al. [26] describe a study carried out using metal-organic chemical vapour deposition (MOCVD) of β-Ga$_2$O$_3$ films. They systematically explore the effects of oxygen:gallium ratio during deposition on the structural and optical properties of the films, and on the growth rate; an optimum ratio for crystal quality of 11.2×10^3 is established. This may be beneficial in a range of device fabrication applications.

In their paper, Sriubas et al. [27] utilise electron beam evaporation to deposit scandia-doped zirconia and scandia-alumina co-doped zirconia thin films. XRD and Raman spectroscopy are used to probe the structure of the films; the addition of aluminium dopant hinders the formation of the cubic zirconia phase while stabilising it at temperatures >300 °C. Relations between substrate temperature, crystallite size and ionic conductivity are presented. This work is likely to be of interest in fuel cell and oxygen sensor development.

Lee et al. [28] investigate a novel variant of CVD as a fabrication method for durable superhydrophobic coatings. This method allows the manufacture of such coatings at relatively low temperatures and negates the requirement for substrate pre-treatment. The applicability to paper and cotton fabric is also demonstrated. This class of material has many potential uses, including in antifouling, water repellents and self-cleaning surfaces.

Stachiv and Gan [29] present a very interesting application of microcantilevers as typically used in conventional atomic force microscopy (AFM), repurposed to facilitate measurement of Young's and shear moduli, Poisson's ratio, and film density. This could have significance in ultrathin film analysis, relevant in photovoltaic, optical, microelectronic and sensor applications.

In their paper, Tillmann et al. [30] discuss a combination of thermal spraying and PVD (magnetron sputtering) for the manufacture of Ni/Ni-20Cr thin film thermocouples for use in process monitoring of flat plastic film extrusion. Polypropylene foils of good surface quality are shown to be producible using such devices, which are also highly stable in operation. This work demonstrates that PVD-deposited thermocouples are a promising approach for automated manufacturing process monitoring.

Zhang et al. [31] investigated the use of an oxygen ion beam to improve the performance of magnesium fluoride (MgF$_2$) as an optical coating layer in the visible, near infrared (NIR) and mid-wavelength infrared (MWIR) spectral regions. They report that as oxygen flow increases, the film density and refractive index increases; an MgO phase is formed within the MgF$_2$ matrix, with oxygen filling F$^-$ ionic vacancies. This hinders the combination of magnesium ions with hydroxyl groups present in the atmosphere, and water adsorption is also decreased and thus reducing optical absorption of the film in the infrared. This also, however, has the undesirable effect of increasing film stress and weakening adhesion to the substrate. To circumvent this issue, the authors propose a solution involving a very thin MgF$_2$ layer being deposited first, without ion assist, followed by a second oxygen ion beam-assisted MgF$_2$ layer. MgF$_2$ is a commonly used material in optical coatings, particularly in anti-reflection coatings, and so this research is likely to be of great interest in the optical coating field.

In their paper, Zhang et al. [32] propose a metal-insulator-metal structure, designed to operate as a dual-function metalens. This metalens has a focal length of 5 μm for x-polarisation and 15 μm for y-polarisation, and functions within the 750–850 nm waveband. The structure may also function as a type of beam splitter. This is likely to be of interest within the field of metamaterials research.

In the area of tribological coatings research, Smolik et al. [33] investigate the effect of tungsten doping of magnetron sputtered titanium diboride coatings. They report improvement of brittleness and fracture toughness of such coatings by addition of tungsten in the range of 0–10% at. This has the effect of changing the coating's microstructure, from a typical columnar structure to a nanocomposite structure, decreasing the energy of individual cracks during fracture testing and consequently increasing the fracture toughness. This could have many potential applications, including in armour manufacture, wear-resistant coatings and cutting tools.

Finally, and staying on the topic of tribological coatings, Skordaris et al. [34] describe the reduction in residual stresses in 5 μm-thick nanocrystalline diamond coatings deposited on cemented carbide insert tools, via an optimised annealing process. This results in an impressive enhancement of fatigue strength and milling performance of such tools, increasing the components' service life by up to four times that of a non-annealed coated tool. This is likely to be of great interest in diamond coating research and industrial cutting applications.

Conflicts of Interest: The authors declare no conflict of interest.

References

1. Papadopoulos, N.; Qiu, W.; Ameys, M.; Smout, S.; Willegems, M.; Deroo, F.; van der Steen, J.; Kronemeijer, A.J.; Dehouwer, M.; Mityashin, A.; et al. Touchscreen tags based on thin-film electronics for the Internet of Everything. *Nat. Electron.* **2019**, *2*, 606–611. [CrossRef] [PubMed]
2. Klee, M.; Beelen, D.; Keurl, W.; Kiewitt, R.; Kumar, B.; Mauczokl, R.; Reimann, K.; Renders, C.; Roest, A.; Roozeboom, F.; et al. Application of Dielectric, Ferroelectric and Piezoelectric Thin Film Devices in Mobile Communication and Medical Systems. In Proceedings of the 15th IEEE International Symposium on the Applications of Ferroelectrics, Sunset Beach, NC, USA, 30 July–2 August 2006; Volume 2006, pp. 9–16. [CrossRef]
3. Fleming, L.; Gibson, D.; Song, S.; Li, C.; Reid, S. Reducing N_2O induced cross-talk in a NDIR CO_2 gas sensor for breath analysis using multilayer thin film optical interference coatings. *Surf. Coat. Technol.* **2018**, *336*, 9–16. [CrossRef]
4. Birney, R.; Steinlechner, J.; Tornasi, Z.; MacFoy, S.; Vine, D.; Bell, A.; Gibson, D.; Hough, J.; Rowan, S.; Sortais, P.; et al. Amorphous Silicon with Extremely Low Absorption: Beating Thermal Noise in Gravitational Astronomy. *Phys. Rev. Lett.* **2018**, *121*, 191101. [CrossRef] [PubMed]
5. Ștefanov, T.; Vardhan, H.; Maraka, R.; Meagher, P.; Rice, J.; Sillekens, W.; Browne, D. Thin film metallic glass broad-spectrum mirror coatings for space telescope applications. *J. Non Cryst. Solids X* **2020**, *7*, 100050. [CrossRef]
6. Craig, K.; Steinlechner, J.; Murray, P.; Bell, A.; Birney, R.; Haughian, K.; Hough, J.; MacLaren, I.; Penn, S.; Reid, S.; et al. Mirror Coating Solution for the Cryogenic Einstein Telescope. *Phys. Rev. Lett.* **2019**, *122*, 231102. [CrossRef]
7. Robertson, S.; Gibson, D.; MacKay, W.; Reid, S.; Williams, C.; Birney, R. Investigation of the antimicrobial properties of modified multilayer diamond-like carbon coatings on 316 stainless steel. *Surf. Coat. Technol.* **2017**, *314*, 72–78. [CrossRef]
8. Schaefer, C.; Bräuer, G.; Szczyrbowski, J. Low emissivity coatings on architectural glass. *Surf. Coat. Technol.* **1997**, *93*, 37–45. [CrossRef]
9. Peumans, P.; Uchida, S.; Forrest, S. Efficient bulk heterojunction photovoltaic cells using small-molecular-weight organic thin films. *Mater. Sustain. Energy* **2010**, 94–98. [CrossRef]
10. Fthenakis, V. Sustainability of photovoltaics: The case for thin-film solar cells. *Renew. Sustain. Energy Rev.* **2009**, *13*, 2746–2750. [CrossRef]
11. Duyar, Ö.; Placido, F.; Durusoy, H.Z. Optimization of TiO_2 films prepared by reactive electron beam evaporation of Ti_3O_5. *J. Phys. D Appl. Phys.* **2008**, *41*, 9. [CrossRef]
12. Song, S.; Keating, M.; Chen, Y.; Placido, F. Reflectance and surface enhanced Raman scattering (SERS) of sculptured silver films deposited at various vapor incident angles. *Meas. Sci. Technol.* **2012**, *23*, 8. [CrossRef]
13. Lappschies, M.; Görtz, B.; Ristau, D. Application of optical broadband monitoring to quasi-rugate filters by ion-beam sputtering. *Appl. Opt.* **2006**, *45*, 1502–1506. [CrossRef] [PubMed]
14. Stewart, A.; Lu, S.; Tehrani, M.; Volk, C. Ion Beam Sputtering of Optical Coatings. In Proceedings of the SPIE 2114, Laser-Induced Damage in Optical Materials, Boulder, CO, USA, 27–29 October 1993. [CrossRef]
15. Birney, R.; Cumming, A.; Campsie, P.; Gibson, D.; Hammond, G.; Hough, J.; Martin, I.; Reid, S.; Rowan, S.; Song, S.; et al. Coatings and surface treatments for future gravitational wave detectors. *Class. Quantum Gravity* **2017**, *34*, 235012. [CrossRef]

16. Larsen, M. Chemical Vapour Deposition of Large Area Graphene. Ph.D. Thesis, Danmarks Tekniske Universitet (DTU Nanotech), Lyngby, Danmark, 2015. Available online: https://orbit.dtu.dk/files/110881784/PhD_thesis_Martin_Benjamin_Larsen.pdf (accessed on 11 December 2020).

17. Mackenzie, D.; Buron, J.; Whelan, P.; Jessen, B.; Silajdźić, A.; Pesquera, A.; Centeno, A.; Zurutuza, A.; Bøggild, P.; Petersen, D. Fabrication of CVD graphene-based devices via laser ablation for wafer-scale characterization. *2D Mater.* **2015**, *2*, 045003. [CrossRef]

18. Faller, F.; Hurrle, A. High-temperature CVD for crystalline-silicon thin-film solar cells. *IEEE Trans. Electron. Devices* **1999**, *46*, 10. [CrossRef]

19. Ponon, N.; Appleby, D.; Arac, E.; King, P.; Ganti, S.; Kwa, K.; O'Neill, A. Effect of deposition conditions and post deposition anneal on reactively sputtered titanium nitride thin films. *Thin Solid Film.* **2015**, *578*, 31–37. [CrossRef]

20. Vajente, G.; Birney, R.; Ananyeva, A.; Angelova, S.; Asselin, R.; Baloukas, B.; Bassiri, R.; Billingsley, G.; Fejer, M.; Gibson, D.; et al. Effect of elevated substrate temperature deposition on the mechanical losses in tantala thin film coatings. *Class. Quantum Gravity* **2018**, *35*, 7. [CrossRef]

21. Mazur, M.; Wojcieszak, D.; Domaradzki, J.; Kaczmarek, D.; Song, S.; Placido, F. TiO_2/SiO_2 multilayer as an antireflective and protective coating deposited by microwave assisted magnetron sputtering. *Opto Electron. Rev.* **2013**, *21*, 233–238. [CrossRef]

22. Johnson, R.; Hultqvist, A.; Bent, S. A brief review of atomic layer deposition: From fundamentals to applications. *Mater. Today* **2014**, *17*, 236–246. [CrossRef]

23. George, S. Atomic Layer Deposition: An Overview. *Chem. Rev.* **2010**, *110*, 111–131. [CrossRef]

24. Hsieh, Y.; Kau, L.; Huang, H.; Lee, C.; Fuh, Y.; Li, T. In Situ Plasma Monitoring of PECVD nc-Si:H Films and the Influence of Dilution Ratio on Structural Evolution. *Coatings* **2018**, *8*, 238. [CrossRef]

25. Schwinger, L.; Lehmann, S.; Zielbauer, L.; Scharfe, B.; Gerdes, T. Aluminium Coated Micro Glass Spheres to Increase the Infrared Reflectance. *Coatings* **2019**, *9*, 187. [CrossRef]

26. Li, Z.; Jiao, T.; Hu, D.; Lv, Y.; Li, W.; Dong, X.; Zhang, Y.; Feng, Z.; Zhang, B. Study on β-Ga_2O_3 Films Grown with Various VI/III Ratios by MOCVD. *Coatings* **2019**, *9*, 281. [CrossRef]

27. Sriubas, M.; Kainbayev, N.; Virbukas, D.; Bočkutė, K.; Rutkūnienė, Ž.; Laukaitis, G. Structure and Conductivity Studies of Scandia and Alumina Doped Zirconia Thin Films. *Coatings* **2019**, *9*, 317. [CrossRef]

28. Lee, H.; Kim, H.; Lee, J.; Kwak, J. Fabrication of a Conjugated Fluoropolymer Film Using One-Step iCVD Process and its Mechanical Durability. *Coatings* **2019**, *9*, 430. [CrossRef]

29. Stachiv, I.; Gan, L. Simple Non-Destructive Method of Ultrathin Film Material Properties and Generated Internal Stress Determination Using Microcantilevers Immersed in Air. *Coatings* **2019**, *9*, 486. [CrossRef]

30. Tillmann, W.; Kokalj, D.; Stangier, D.; Schöppner, V.; Malatyali, H. Combining Thermal Spraying and Magnetron Sputtering for the Development of Ni/Ni-20Cr Thin Film Thermocouples for Plastic Flat Film Extrusion Processes. *Coatings* **2019**, *9*, 603. [CrossRef]

31. Zhang, G.; Fu, X.; Song, S.; Guo, K.; Zhang, J. Influences of Oxygen Ion Beam on the Properties of Magnesium Fluoride Thin Film Deposited Using Electron Beam Evaporation Deposition. *Coatings* **2019**, *9*, 834. [CrossRef]

32. Zhang, Y.; Yang, B.; Liu, Z.; Fu, Y. Polarization Controlled Dual Functional Reflective Planar Metalens in Near Infrared Regime. *Coatings* **2020**, *10*, 389. [CrossRef]

33. Smolik, J.; Kacprzyńska-Gołacka, J.; Sowa, S.; Piasek, A. The Analysis of Resistance to Brittle Cracking of Tungsten Doped TiB_2 Coatings Obtained by Magnetron Sputtering. *Coatings* **2020**, *10*, 807. [CrossRef]

34. Skordaris, G.; Kotsanis, T.; Boumpakis, A.; Stergioudi, F. Improvement of the Interfacial Fatigue Strength and Milling Behavior of Diamond Coated Tools via Appropriate Annealing. *Coatings* **2020**, *10*, 821. [CrossRef]

Publisher's Note: MDPI stays neutral with regard to jurisdictional claims in published maps and institutional affiliations.

Article

In Situ Plasma Monitoring of PECVD nc-Si:H Films and the Influence of Dilution Ratio on Structural Evolution

Yu-Lin Hsieh [1], Li-Han Kau [2], Hung-Jui Huang [2], Chien-Chieh Lee [3], Yiin-Kuen Fuh [1,*] and Tomi T. Li [1]

[1] Department of Mechanical Engineering, National Central University, Taoyuan 32001, Taiwan; firenze114@gmail.com (Y.-L.H.); tomili@cc.ncu.edu.tw (T.T.L.)
[2] Opto-mechatronics Engineering, National Central University, Taoyuan 32001 Taiwan; as811224@gmail.com (L.-H.K.); rayan811029@gmail.com (H.-J.H.)
[3] Optical Science Center, National Central University, Taoyuan 32001 Taiwan; jjlee@ncu.edu.tw
* Correspondence: mikefuh@cc.ncu.edu.tw; Tel.: +886-34227151 (ext. 34305)

Received: 12 April 2018; Accepted: 2 July 2018; Published: 6 July 2018

Abstract: We report plasma-enhanced chemical vapor deposition (PECVD) hydrogenated nano-crystalline silicon (nc-Si:H) thin films. In particular, the effect of hydrogen dilution ratio ($R = H_2/SiH_4$) on structural and optical evolutions of the deposited nc-Si:H films were systematically investigated including Raman spectroscopy, Fourier-transform infrared spectroscopy (FTIR) and low angle X-ray diffraction spectroscopy (XRD). Measurement results revealed that the nc-Si:H structural evolution, primarily the transition of nano-crystallization from the amorphous state to the nanocrystalline state, can be carefully induced by the adjustment of hydrogen dilution ratio (R). In addition, an in situ plasma diagnostic tool of optical emission spectroscopy (OES) was used to further characterize the crystallization rate index ($H_\alpha*/SiH*$) that increases when hydrogen dilution ratio (R) rises, whereas the deposition rate decreases. Another in situ plasma diagnostic tool of quadruple mass spectrometry (QMS) also confirmed that the "optimal" range of hydrogen dilution ratio (R = 30–40) can yield nano-crystalline silicon (n-Si:H) growth due to the depletion of higher silane radicals. A good correlation between the plasma characteristics by in situ OES/QMS and the film characteristics by XRD, Raman and FTIR, for the transition of a-Si:H to nc-Si:H film from the hydrogen dilution ratio, was obtained.

Keywords: PECVD; plasma diagnostics; nc-Si:H; RF-PECVD; Fourier-transform infrared spectroscopy (FTIR); quadruple mass spectrometry (QMS); optical emission spectroscopy (OES); X-ray diffraction spectroscopy (XRD)

---

1. Introduction

Hydrogenated nanocrystalline silicon (nc-Si:H) is a promising material to have distinguishing characteristics of higher carrier mobility, greater doping efficiency and efficient visible photoluminescence [1–3]. Generally, the re-crystallization technique such as rapid thermal annealing [4], excimer laser [5], and aluminium induced crystallization [6] were typically used to obtain the nc-Si:H films with a variety of applications in silicon-based tandem device solar cells [7,8], photovoltaic and optoelectronic devices [9].

On the other hand, several chemical vapor deposition (CVD) techniques have been widely used to deposit the thin films include hot wire CVD (HW-CVD) [10], layer-by-layer (LBL) deposition [11], magnetron sputtering [12], electron cyclotron resonance CVD [13], and plasma enhanced CVD (PECVD) [11] and its variant, very high frequency plasma-enhanced chemical vapor deposition

(VHF-PECVD) [14]. In particular, PECVD is the one of the standard industrial processes in the solar cells and Thin-Film Transistor (TFT) technology.

In the PECVD growth of thin films, critical process conditions, including the gas flow composition, chamber environment, RF power density [15,16], process gas mixtures of hydrogen (H_2) and silane (SiH_4) [17,18], should be carefully monitored. Therefore, plasma diagnostics is essentially useful for investigations of thin film growth plasma conditions. Several diagnostic tools such as optical emission spectroscopy (OES) [19,20] and quadrupole mass spectrometry (QMS) [21–23] were previously reported to in situ monitor the plasma chemistry. The deposited thin films can be further characterized by Fourier-transform infra-red spectroscopy (FTIR) [24,25], X-ray diffraction spectroscopy (XRD) and Raman spectroscopy such that the structure of the materials can be extensively investigated. For the low temperature deposition of polycrystalline silicon and solar cell applications, a thorough review had been reported and the characteristics of deposition and physical properties were comprehensively indicated to be the guiding parameters for achieving poly Si films at high deposition rates [26].

In this paper, we present the effect of hydrogen dilution ratio ($R = H_2/SiH_4$) on structural and optical properties of PECVD nc-Si:H films. In addition, Raman spectroscopy and XRD tools are performed to study the growth of a-Si:H transition to nanocrystalline silicon (nc-Si:H). By varying the hydrogen dilution ratio, the film properties were experimentally correlated with the emission intensities of the excited radicals from plasma diagnostic tools of OES and QMS. The obtained results demonstrated that hydrogen dilution ratio (R) played a critical role in PECVD nc-Si:H films.

2. Experimental

Intrinsic hydrogenated nanocrystalline silicon (nc-Si:H) films were PECVD deposited with silane and hydrogen which were used as a source gas and diluent, respectively. The main purpose of this study aims to quantatively characterize the amorphous to nano-crystalline transition, therefore, a systematic variation of hydrogen dilution ratio ($R = H_2/SiH_4$) of 5–40 was carried out with RF power frequency 13.56 MHz and substrate temperature (210 °C). For characterization purpose, films were deposited on 2×2 cm^2 Cz(100) n-type 1–5 Ωcm^2 single-side polished wafer and Corning eagle XG glass (Corning Incorporated, Corning, NY, USA). Prior to deposition, the wafers were cleaned using H_2O_2:H_2SO_4 = 1:2 solution for 5 min, followed by dipping in 2% HF for 1 min to remove native oxide, rinsed in DI water, dried in N_2 atmosphere and then the wafers were immediately transferred into vacuum chamber to prevent wafers from the native oxide layer growth.

Raman spectroscopy (Horiba iHR550, Irvine, CA, USA), Fourier-transform infrared spectroscopy (FTIR-Perkin Elmer Spectrum 100, Shelton, CT, USA), Quadruple mass spectrometry (QMS-Hiden PSM003P, Warrington, United Kingdom), Optical emission spectroscopy (OES-ocean optics usb2000+, Winter Park, FL, USA), X-ray diffraction spectroscopy (XRD-PANalytical Empyrean, Royston, United Kingdom) measurements were performed. The film thicknesses were measured using a profilometer of Alpha-step (KLA-Tencor D-300, Milpitas, CA, USA). OES measurement was primarily used to investigate the in situ excited radicals inside plasma. QMS is adopted to quantitatively monitor the plasma condition in assisting the development of deposition process.

Films were deposited in a commercially available PECVD unit (Creating Nano-Technologies, PE-001, Tainan, Taiwan) as schematically shown in Figure 1. This parallel plate PECVD reactor with two attached plasma in situ diagnostics of OES and QMS on the side is used. The in situ QMS was applied and the position of the QMS orifice is fixed to the chamber wall quartz window. Threshold ionization mass spectrometry (TIMS) was also used to detect atomic mass unit (amu) in the range of 0–90 and the time resolution is set at ~240 s sweeping time during measurements. The process chamber was kept at a base pressure less than 10^{-6} Torr via the coupled turbo molecular pump (TMP). The detailed deposition parameters are illustrated in Table 1.

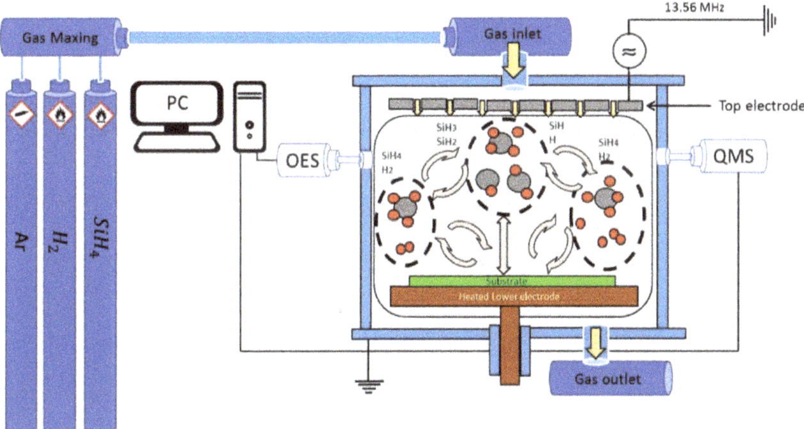

Figure 1. Schematic diagram of the PECVD equipment with a 13.56 MHz radio frequency (RF) and in situ plasma diagnostic tool of OES and QMS on the side. H_2 gas, SiH_4 gas and Ar gas are introduced in the chamber from upstream of the plasma zone.

Table 1. Main process parameters used of the deposition of intrinsic nc-Si:H thin films by PECVD.

Source Gases	Pressure	Power	Gas Flow Ratio			Deposition Time	Distance	Substrate Temperature
			(1) Ar	(2) SiH$_4$	(3) H$_2$			
Ar, SiH$_4$, H$_2$	300 mTorr	100 W	3 sccm	5 sccm	25, 50, 100, 150, 200 sccm	60 min	25 mm	210 °C

3. Results and Discussion

3.1. Effect of the Hydrogen Dilution Ratio on the Deposition Rate

Figure 2 shows the relationship between the deposition rate (black circle line), SiH*/Ar and H_α*/Ar ratio as extrapolated from optical emission spectra as a function of different hydrogen dilution ratio ($R = H_2/SiH_4$). In Figure 2, the intensity of the H_α*/Ar ratio is comparatively higher than that of the SiH*/Ar counterpart for the hydrogen dilution ratio (R), which is greater than 20. It was experimentally observed that the deposition rate decreases with the increase in hydrogen dilution ratio (R). Similar trends between the SiH* intensity and the deposition rate were reported [27], and, furthermore, ion bombardment [28] and hydrogen etch (H_α*/Ar ratio) [29] impact will result in heavy defects and easily crumbled film. Thus, the deposition rate decreases as hydrogen dilution ratio increases, as shown in Figure 2.

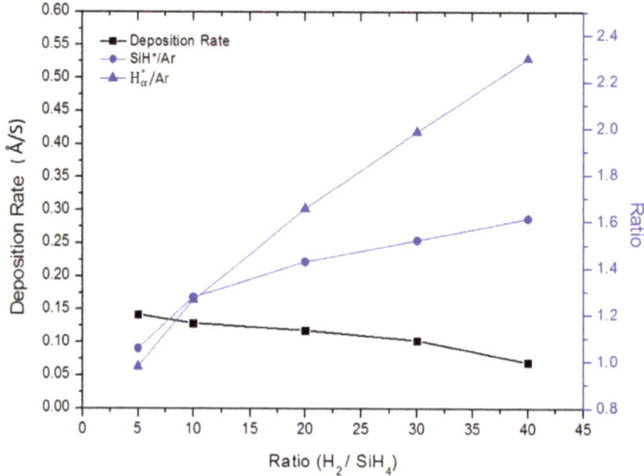

Figure 2. Effect of the hydrogen dilution ratio (*R*) on the deposition rate (black line) the SiH* emission intensity (blue circle line) and H$_\alpha$* emission intensity (blue triangle line).

3.2. Low Angle XRD Analysis

XRD patterns of five typical samples under the same RF power (100 W) but different hydrogen dilution ratios are presented in Figure 3. The transformation from the amorphous to the partially crystalline state can be observed for peaks at angles 28.5°, 47.2° and 56.2°, as assigned to Si(111), Si(221) and Si(311) reflection planes of faced-centered cubic silicon, respectively.

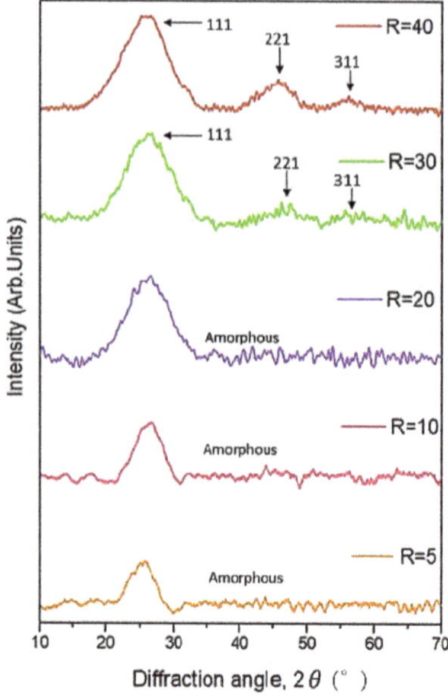

Figure 3. Low angle XRD pattern of nc-Si:H films deposited under different hydrogen dilution ratio (*R*) using the PECVD method.

In the case of low hydrogen dilution ratio ($R = 5$), no crystal grains and apparent structural evolution were detected in the a-Si:H thin films. Further increasing the hydrogen dilution ratio ($R = 10$ and 20) still shows amorphous films centered at 2θ ~27.5°. On the other hand, films deposited with higher hydrogen dilution ratio ($R = 30$ and 40) the XRD pattern show three peaks at 2θ ~27.3°, 46.8° and 56.5°, respectively, vividly showing the existence of nanocrystalline-Si phase. This results are in good agreement with the other XRD patterns of three peaks at 2θ ~28.4°, 47.5° and 56.1° [30]. The higher the dilution ratio R, the crystal grain size increases (as evidenced from Raman spectra) with higher crystallinity (the volume fraction of partially nanocrystalline structure) with the increase of the hydrogen dilution ratio. Therefore, it is concluded that the hydrogen dilution ratio in PECVD is a critical processing factor for the growth of nc-Si:H films.

3.3. Raman Spectroscopy Analysis

Raman spectroscopy is used to investigate the nanocrystalline structure of deposited films. Figure 4 shows the Raman spectra of the nc-Si:H films deposited at different hydrogen dilution ratio (R) ranging from 5 to 40. The volume fraction of crystallites (X_{Raman}) and crystallite size (d_{Raman}) of nc-Si:H films can be calculated using the Levenberg–Marquardt method by de-convoluting into three peaks as a crystalline peak (~510 cm^{-1}), an amorphous peak (~475 cm^{-1}) and an intermediate peak (~490 cm^{-1}), respectively [31]. Typical de-convoluted Raman spectra for nc-Si:H film deposited at $R = 40$ are shown in Figure 5. The crystalline fraction (X_{Raman}) can be calculated [31]:

$$X_{Raman} = \frac{I_C + I_m}{I_C + I_m + I_a} \tag{1}$$

where I_C, I_m and I_a are the integrated intensity of 520 cm^{-1} (the crystalline phase), 500 cm^{-1} (the intermediate phase) and 480 cm^{-1} (the amorphous phase). The crystallite size (d_{Raman}) can also be calculated:

$$d_{Raman} = 2\pi\sqrt{\frac{\beta}{\Delta\omega}} \tag{2}$$

where $\Delta\omega$ is the peak shift compared to c–Si peak located ~520 cm^{-1} and $\beta = 2.0$ cm^{-1} nm^2 [8].

At a low hydrogen dilution ratio ($R = 5$, 10 and 20) as seen in Figure 4, the deposited films show the typical a-Si:H absorption peak centered ~480 cm^{-1}. However, the onset of nanocrystallization can be experimentally observed of the film deposited at a hydrogen dilution ratio ($R = 30$), which corresponds to the Transverse Optic (TO) phonon peak centred −507 cm^{-1} of the nanocrystalline phase [32]. For this film, X_{Raman} is calculated ~86.13% and d_{Raman} is ~2.46 nm. Further increasing in hydrogen dilution ratio ($R = 40$), the peak shifts towards ~509 cm^{-1} which indicates the increase in the volume fraction of crystallites and its size. For this film, X_{Raman} is ~89.25% and d_{Raman} is ~2.68 nm. The increase in hydrogen dilution ratio (R) in PECVD results in an amorphous-to-nanocrystalline transition in the film. Previously reported similar results for nc-Si:H films deposited by using layer-by-layer (LBL) deposition [33] and increasing RF power [30] concurrently promotes the hydrogen etching and nano-crystallization in the films. Therefore, Raman spectroscopy analysis indicates that the critical process parameter includes both hydrogen dilution ratio and RF power in PECVD, which are functionally equivalent to induce nc-Si:H films. The Raman spectra of the amorphous Si films for the peak at 480 cm^{-1} for the dilution ratio ($R = 5$, 10 and 20) were measured and the full width at half maximum (FWHM) was calculated as 46.8 cm^{-1}, 48.4 cm^{-1}, and 54.3 cm^{-1}, respectively. The above measurement is in agreement with the reported Raman spectra [34] of the amorphous Si films such that the FWHM is around 53 ± 7 cm^{-1}.

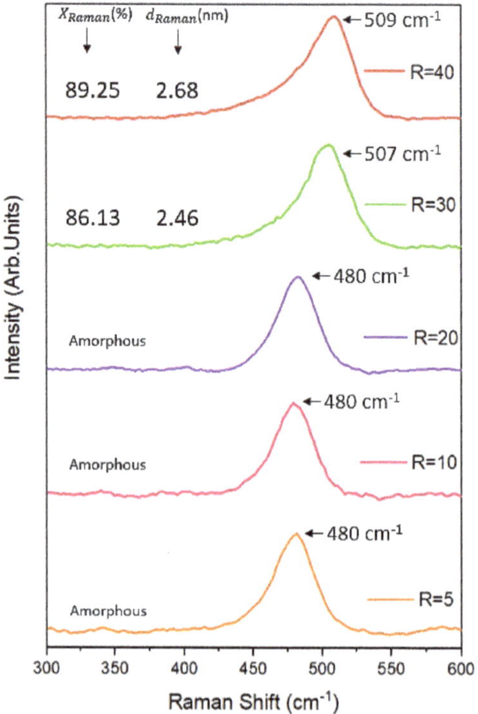

Figure 4. Raman spectra of nc-Si:H films deposited as a function of different hydrogen dilution ratio (*R*) using the PECVD method.

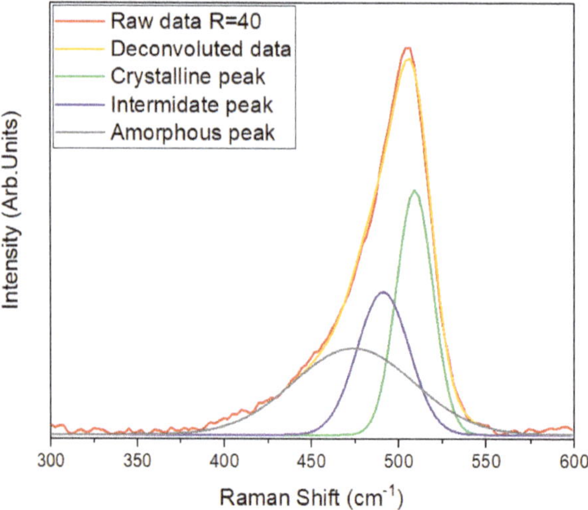

Figure 5. Typical Gaussian de-convoluted Raman spectra of nc-Si:H films deposited at hydrogen dilution ratio (*R* = 40) using PECVD; Raman spectra of the nano-crystalline silicon intrinsic layers (yellow line), and the red line is fitted lines from three Gaussian peaks at ~475, ~490, and ~510 cm^{-1}.

3.4. Fourier Transform Infra-Red (FTIR) Spectroscopy Analysis

In order to investigate the hydrogen bonding and resultant total hydrogen content in the nc-Si:H film, FTIR spectra of different dilution ratios are performed in Figure 6. It can be measured that two major absorption bands centered at ~620 and ~2000 cm^{-1} (wagging/stretching modes, respectively), which are closely related to the vibrations of mono-hydrogen (Si–H) bonded species [35]. The spectra also exhibit an absorption peak centered at ~1060 cm^{-1}, which can be Si–O–Si stretching vibrations of a typical undoped nc-Si:H thin films [36]. In addition, a lesser intensity peak centered at ~885 cm^{-1} has been observed as the bending vibrational modes of Si–H$_2$ [37]. With the increase in hydrogen dilution ratio (R), the intensity of absorption band at ~620 cm^{-1} also incrementally increases. In summary, the increase of hydrogen dilution ratio (R) will promote the increased intensity at ~620 cm^{-1} while as absorption band centered at ~2000 cm^{-1} shifts incrementally toward the peak at 2100 cm^{-1}. The di-hydride (Si–H$_2$) and poly-hydride (Si–H$_2$)$_n$ bonded species can be assigned as ~2100 cm^{-1} stretching vibrational modes [38]. The experimental results indicate that, with an increase in hydrogen dilution ratio (R), the absorption peak shifts from Si–H to Si–H$_2$ and (Si–H$_2$)$_n$ bonded species in films. In addition, it has been reported that the integrated intensity of the peak at 620 cm^{-1} is the best reliable measure of hydrogen content (C_H) [39] by taking the oscillator strength value [37].

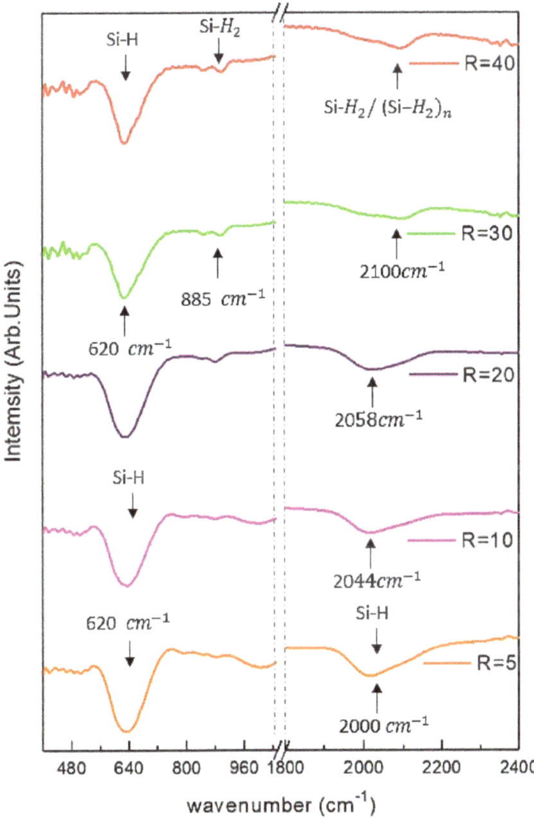

Figure 6. FTIR absorbance spectra of nc-Si:H films (at normalized thickness) deposited by PECVD at different hydrogen dilution ratios (R). The signal located 620 cm^{-1} and 2100 cm^{-1} are assigned as the wagging/rocking and stretching mode of nc–Si–H bond, respectively (see the text).

Figure 7 shows the relationship between the hydrogen content (C_H) and different hydrogen dilution ratio (R). It is experimentally demonstrated that hydrogen content initially increases with the increase in dilution ratio (R) from 10 to 20, then drops abruptly in the R range of 20–30, and follows the gradually increasing hydrogen content as R values from 30 to 40. A previously reported hydrogen rich study indicated that an increase in RF power promotes the silane dissociation and produces more SiH_x (x = 1, 2, 3) radicals [40] as well as the increased electron temperature [41]. In our study, RF power is set at 100 W, while the hydrogen dilution ratio (R) shows the similar impact such that the hydrogen rich precursors have higher sticking probability as the hydrogen content increases. A similar trend also indicates that an increase in hydrogen dilution ratio (R) also boosts the density of atomic H in the deposition chamber and further enhances the surface diffusion with a hydrogen-accumulated growing surface [42]. Both the synergetic effect of the hydrogen etching effect [43] and amorphous-to-nanocrystalline transition can be observed in a hydrogen dilution ratio (R) range of 30–40, which is in agreement with the Raman shift in Figure 4.

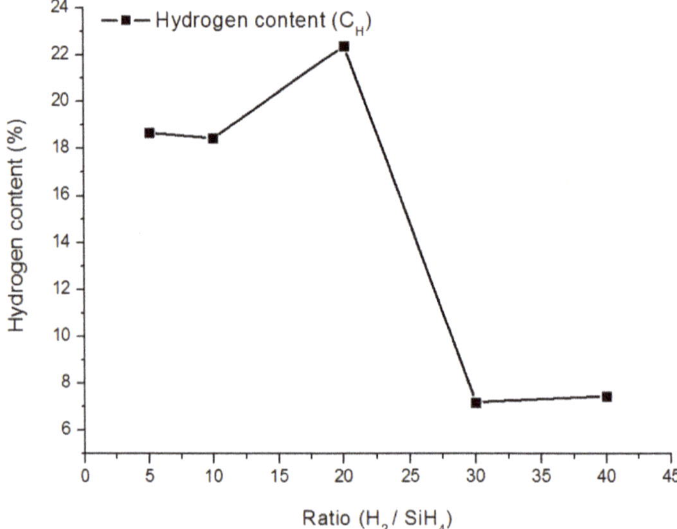

Figure 7. Variation of hydrogen content as a function of hydrogen dilution ratio (R) for nc-Si:H films.

3.5. Optical Emission Spectra (OES) Spectroscopy Analysis

Figure 8a shows a series of OES of RF glow discharges of H_α^*/SiH^* and Si^*/SiH^* with a different dilution ratio (R). The main characteristic spectra of excited radicals correspond to the emission of SiH^* (414.30 nm), H_α (486.10 nm), H_β (656.30 nm), Ar^* (750.30 nm) and O^* (844.30 nm), which is directly correlated with the SiH_4 electron collision and dissociative excitation. Spatially resolved OES measurements can be used to record the SiH radicals in axial concentration profiles. Plots of crystallization rate indexes (H_α^*/SiH^*) and electron temperature indexes (Si^*/SiH^*) versus dilution ratio (R) are shown in Figure 8b. The result shows that the crystallization rate index (H_α^*/SiH^*) increases when the dilution ratio rises, resulting in the opposite trend of the deposition rate. On the other hand, the electron temperature (Si^*/SiH^*) is increasing along with a dilution ratio (R) increase, causing increased ion bombardment [29] and thus decreasing the deposition rate.

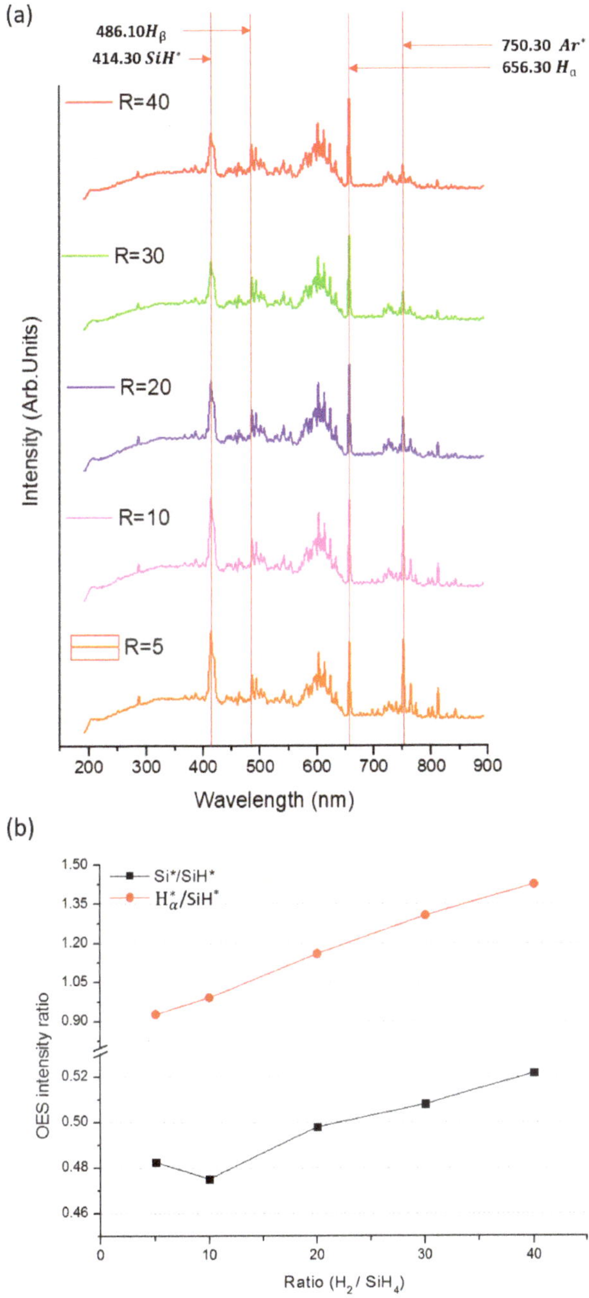

Figure 8. (**a**) Optical emission spectra measured during the deposition of nc-Si:H thin film as a function of hydrogen dilution ratio (*R*). Some prominent optical emission peaks and band structures for silane–hydrogen plasma are labelled; (**b**) OES intensity ratio of H_α^*/SiH* (red line) and ratio of electron temperature Si*/SiH* (black line) calculated during the deposition of nc-Si:H thin film as a function of varying hydrogen dilution ratio (*R*).

3.6. Quadruple Mass Spectra (QMS) Spectroscopy Analysis

Measured results of QMS during nc-Si:H deposition in Figure 9 indicates some distinguishably identified fingerprints of the radicals SiH_x^+ ($0 < x < 4$). The TIMS method shows the trends in relative densities of four mono-silane radicals (SiH_3^+, SiH_2^+, SiH^+ and Si^+) by varying the hydrogen dilution ratio in argon–silane–hydrogen plasmas. PECVD nc-Si:H deposition shows that a critical point of relative densities is obtained as the SiH^+ radical reached a threshold value at R ~30 (SiH_3^+ radicals remain almost unchanged, pink bar). This study shows significantly different plasma composition at different silane dilution in hydrogen (R = 5–40), validating that SiH^+ is the dominant radical in the growth of nc-Si:H layers, [44]. Therefore, in the range of R = 30 and 40 of high silane dilution, QMS experiments confirmed that nano-crystalline silicon (nc-Si:H) growth is closely related to the depletion of higher silane radicals [42,45].

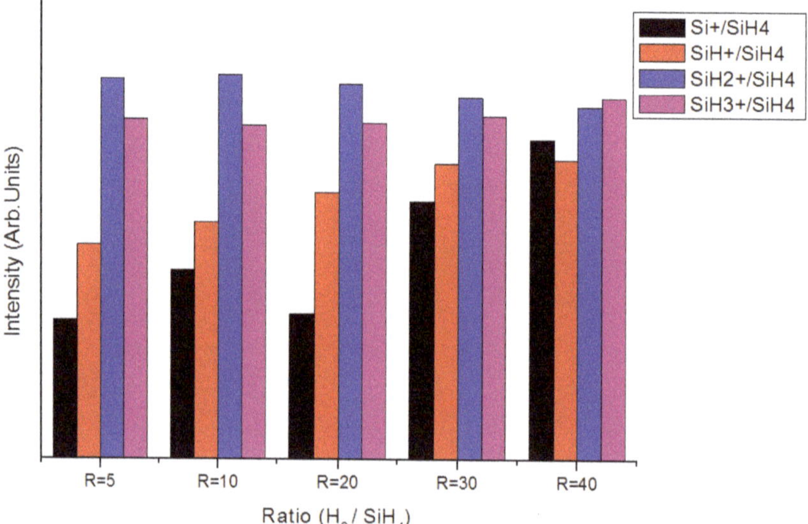

Figure 9. Relative density QMS measured during typical nc-Si:H deposition as a function of varying hydrogen dilution ratios (R).

4. Conclusions

Hydrogenated nanocrystalline silicon (nc-Si:H) thin films were grown by the PECVD system. The results show that as hydrogen dilution ratio (R) increases, the deposition rate decreases. It was also revealed by low angle XRD and Raman spectroscopy that nc-Si:H films can be induced via the adjustment of hydrogen dilution ratio (R). Moreover, FTIR spectroscopy indicates that the shifts of predominant hydrogen bonding (from Si–H to Si–H_2 and (Si–H_2)$_n$ bonded species) are closely related to the increase in the hydrogen dilution ratio (R). At the dilution ratio (R = 40), nc-Si:H films have a crystallite size of ~2.3 nm with a degree of crystallinity and ~89.2% were obtained with a low hydrogen content (C_H = 7.45%) at moderately low deposition rate (0.69 Å/s). Furthermore, the QMS plasma diagnostic tool also confirmed the existence of a direct correlation between nc-Si:H film growth and the SiH^+ radicals film, via the tuning of a hydrogen dilution ratio in the range of 5–40. Overall, this study shows a good and important correlation of plasma characteristics by in situ OES/QMS and the end resulted in film characteristics by XRD, Raman and FTIR, for the transition of a-Si:H to nc-Si:H film by a hydrogen dilution ratio.

Author Contributions: Conceptualization, T.T.L., Y.-K.F. and C.-C.L.; Methodology, Y.-L.H.; Software, L.-H.K. and H.-J.H.; Validation, Y.-L.H., L.-H.K. and H.-J.H.; Formal Analysis, Y.-L.H., L.-H.K. and H.-J.H.; Investigation, T.T.L. and Y.-K.F.; Curation, T.T.L, Y.-K.F., C.-C.L. and Y.-L.H.; Writing-Original Draft Preparation, Y.-K.F., T.T.L., L.-H.K. and H.-J.H.; Writing-Review & Editing, Y.-K.F., T.T.L., L.-H.K. and H.-J.H.; Visualization, L.-H.K. and H.-J.H.; Supervision, T.T.L., Y.-K.F. and C.-C.L.; Project Administration, T.T.L., Y.-K.F. and C.-C.L.

Funding: This research was financially funded by the Delta Electronics, Inc., Taiwan under the grant number 10613082.

Acknowledgments: This study was financially supported by Delta Electronics, Inc., Taiwan and Department of Mechanical Engineering, Optical Science Center and Department of Optics and Photonics, National Central University, Taiwan.

Conflicts of Interest: The authors declare no conflict of interest.

References

1.   Zhou, H.P.; Wei, D.Y.; Xu, S.; Xiao, S.Q.; Xu, L.X.; Huang, S.Y.; Guo, Y.N.; Yan, W.S.; Xu, M. Crystalline silicon surface passivation by intrinsic silicon thin films deposited by low-frequency inductively coupled plasma. *J. Appl. Phys.* **2012**, *112*, 013708. [CrossRef]

2.   Tong, G.B.; Aspanut, Z.; Muhamad, M.R.; Rahman, S.A. Optical properties and crystallinity of hydrogenated nanocrystalline silicon (*nc*-Si: H) thin films deposited by RF-PECVD. *Vacuum* **2012**, *86*, 1195–1202. [CrossRef]

3.   Dutta, P.; Paul, S.; Galipeau, D.; Bommisetty, V. Effect of hydrogen plasma treatment on the surface morphology, microstructure and electronic transport properties of *nc*-Si:H. *Thin Solid Films* **2010**, *518*, 6811–6817. [CrossRef]

4.   Zhang, L.; Shen, H.L.; Jiang, X.F.; Qian, B.; Han, Z.D.; Hou, H.H. Influence of annealing temperature on the properties of polycrystalline silicon films formed by rapid thermal annealing of a-Si:H films. *J. Mater. Sci. Mater. Electron.* **2013**, *24*, 4209–4212. [CrossRef]

5.   Adikaari, A.A.D.T.; Mudugamuwa, N.K.; Silva, S.R.P. Nanocrystalline silicon solar cells from excimer laser crystallization of amorphous silicon. *Sol. Energy Mater. Sol. Cells* **2008**, *92*, 634–638. [CrossRef]

6.   Shim, J.H.; Im, S.; Kim, Y.J.; Cho, N.H. Nanostructural and optical features of hydrogenated nanocrystalline silicon films prepared by aluminium-induced crystallization. *Thin Solid Films* **2006**, *503*, 55–59. [CrossRef]

7.   Kaneko, T.; Wakagi, M.; Onisawa, K.; Minemura, T. Change in crystalline morphologies of polycrystalline silicon films prepared by radio-frequency plasma-enhanced chemical vapor deposition using $SiF_4 + H_2$ gas mixture at 350 °C. *Appl. Phys. Lett.* **1994**, *64*, 1865. [CrossRef]

8.   He, Y.; Yin, C.; Cheng, G.; Wang, L.; Liu, X.; Hu, G.Y. The structure and properties of nanosize crystalline silicon films. *J. Appl. Phys.* **1994**, *75*, 797. [CrossRef]

9.   Banerjee, A.; Liu, F.S.; Beglau, D.; Tining, S.; Pietka, G.; Yang, J.; Guha, S. 12.0% Efficiency on Large-Area, Encapsulated, Multijunction nc-Si:H-Based Solar Cells. *IEEE J. Photovolt.* **2012**, *2*, 104–108. [CrossRef]

10. Bakr, N.A.; Funde, A.M.; Waman, V.S.; Kamble, M.M.; Hawaldar, R.R.; Amalnerkar, D.P.; Sathe, V.G.; Gosavi, S.W.; Jadkar, S.R. Influence of deposition pressure on structural, optical and electrical properties of nc-Si:H films deposited by HW-CVD. *J. Phys. Chem. Solids* **2011**, *72*, 685–691. [CrossRef]

11. Goh, B.T.; Wah, C.K.; Aspanut, Z.; Rahman, S.A. Structural and optical properties of *nc*-Si:H thin films deposited by layer-by-layer technique. *J. Mater. Sci. Mater. Electron.* **2014**, *25*, 286–296. [CrossRef]

12. He, H.; Ye, C.; Wang, X.; Huang, F.; Liu, Y. Effect of driving frequency on growth and structure of silicon films deposited by radio-frequency and very-High-frequency magnetron sputtering. *ECS J. Solid State Sci. Technol.* **2014**, *3*, Q74–Q78. [CrossRef]

13. Lee, S.E.; Park, Y.C. Highly-conductive B-doped *nc*-Si:H thin films deposited at room temperature by using SLAN ECR-PECVD. *J. Korean Phys. Soc.* **2014**, *65*, 651–656. [CrossRef]

14. Peng, S.; Wang, D.; Yang, F.; Wang, Z.; Ma, F. Grown Low-Temperature Microcrystalline Silicon Thin Film by VHF PECVD for Thin Films Solar Cell. *J. Nanomater.* **2015**, *2015*, 327596. [CrossRef]

15. Brodsky, M.; Cardona, M.; Cuomo, J.J. Infrared and Raman spectra of the silicon-hydrogen bonds in amorphous silicon prepared by glow discharge and sputtering. *Phys. Rev. B* **1977**, *16*, 3556. [CrossRef]

16. Tauc, J. Absorption edge and internal electric fields in amorphous semiconductors. *Mater. Res. Bull.* **1970**, *5*, 721–729. [CrossRef]

17. Hsieh, Y.L.; Lee, C.C.; Lu, C.C.; Fuh, Y.K.; Chang, J.Y.; Lee, J.Y.; Li, T.T. Structural and electrical investigations of a-Si:H(i) and a-Si:H(n$^+$) stacked layers for improving the interface and passivation qualities. *J. Photonics Energy* **2017**, *7*, 035503. [CrossRef]

18. Dushaq, G.; Nayfeh, A.; Rasras, M. Tuning the optical properties of RF-PECVD grown μc-Si:H thin films using different hydrogen flow rate. *Superlattice Microstruct.* **2017**, *107*, 172–177. [CrossRef]

19. Mataras, D.; Cavadias, S.; Rapakoulias, D. Spatial profiles of reactive intermediates in RF silane discharges. *J. Appl. Phys.* **1989**, *66*, 119. [CrossRef]

20. Dingemans, G.; van den Donker, M.N.; Gordijn, A.; Kessels, W.M.M.; van de Sanden, M.C.M. Probing the phase composition of silicon films in situ by etch product detection. *Appl. Phys. Lett.* **2007**, *91*, 161902. [CrossRef]

21. Fantz, U. Spectroscopic diagnostics and modelling of silane microwave plasmas. *Plasma Phys. Control. Fusion* **1998**, *40*, 1035. [CrossRef]

22. Lisovskiy, V.; Booth, J.-P.; Landry, K.; Douai, D.; Cassagne, V.; Yegorenkov, V. RF discharge dissociative mode in NF$_3$ and SiH$_4$. *Appl. Phys. D* **2007**, *40*, 6631. [CrossRef]

23. Nunomura, S.; Yoshida, I.; Kondo, M. Time-dependent gas phase kinetics in a hydrogen diluted silane plasma. *Appl. Phys. Lett.* **2009**, *94*, 071502. [CrossRef]

24. Strahm, B.; Howling, A.A.; Sansonnens, L.; Hollenstein, C. Plasma silane concentration as a determining factor for the transition from amorphous to microcrystalline silicon in SiH$_4$/H$_2$ discharges. *Plasma Sources Sci. Technol.* **2007**, *16*, 80. [CrossRef]

25. Howling, A.; Strahm, B.; Hollenstein, C. Non-intrusive plasma diagnostics for the deposition of large area thin film silicon. *Thin Solid Films* **2009**, *517*, 6218–6224. [CrossRef]

26. Rath, J.K. Low temperature polycrystalline silicon: A review on deposition, physical properties and solar cell applications. *Sol. Energy Mater. Sol. Cells* **2003**, *76*, 431–487. [CrossRef]

27. Jellison, J.G.E.; Modine, F.A. Parameterization of the optical functions of amorphous materials in the interband region. *Appl. Phys. Lett.* **1996**, *69*, 371. [CrossRef]

28. Lin, L.J.H.; Chiou, Y.-P. Improving thin-film crystalline silicon solar cell efficiency with back surface field layer and blaze diffractive grating. *Sol. Energy* **2012**, *86*, 1485–1490. [CrossRef]

29. Matsuda, A. Growth mechanism of microcrystalline silicon obtained from reactive plasmas. *Thin Solid Films* **1999**, *337*, 1–6. [CrossRef]

30. Jadhavar, A.; Pawbake, A.; Waykar, R.; Waman, V.I.; Rondiya, S.; Shinde, O.; Kulkarni, R.; Rokade, A.; Bhorde, A.; Funde, A.; et al. Influence of RF power on structural optical and electrical properties of hydrogenated nano-crystalline silicon (nc-Si:H) thin films deposited by PE-CVD. *J. Mater. Sci. Mater. Electron.* **2016**, *27*, 12365–12373. [CrossRef]

31. Marquardt, D. An Algorithm for Least-Squares Estimation of Nonlinear Parameters. *J. Soc. Ind. Appl. Math.* **1963**, *11*, 431–441. [CrossRef]

32. Li, Z.; Li, W.; Jiang, Y.; Cai, H.; Gong, Y.; He, J. Raman characterization of the structural evolution in amorphous and partially nanocrystalline hydrogenated silicon thin films prepared by PECVD. *J. Raman Spectrosc.* **2011**, *42*, 415–421. [CrossRef]

33. Tong, G.B.; Muhamad, M.R.; Rahman, S.A. Effects of RF Power on structural properties of nc-Si:H thin films deposited by layer-by-layer (LbL) deposition Technique. *Sains Malays.* **2012**, *41*, 993–1000.

34. Droz, C.; Vallat-Sauvain, E.; Bailat, J.; Feitknecht, L.; Meier, J.; Shah, A. Relationship between Raman crystallinity and open-circuit voltage in microcrystalline silicon solar cells. *Sol. Energy Mater. Sol. Cells* **2004**, *81*, 61–71. [CrossRef]

35. Lucovsky, G. Vibrational spectroscopy of hydrogenated amorphous silicon alloys. *Sol. Cells* **1980**, *2*, 431–442. [CrossRef]

36. Halindintwali, S.; Knoesen, D.; Swanepoel, R.; Julies, B.; Arendse, C.; Muller, T.; Theron, C.; Gordijn, A.; Bronsveld, P.; Rath, J.K.; et al. Improved stability of intrinsic nanocrystalline Si thin films deposited by hot-wire chemical vapour deposition technique. *Thin Solid Films* **2007**, *515*, 8040–8044. [CrossRef]

37. Knights, J.C.; Lucovsky, G.; Nemanich, R.J. Defects in plasma-deposited a-Si:H. *J. Non-Cryst. Solids* **1979**, *32*, 393–403. [CrossRef]

38. Tsu, D.; Lucovsky, G.; Dadison, B. Effects of the nearest neighbors and the alloy matrix on SiH stretching vibrations in the amorphous SiO$_r$:H (0 < r < 2) alloy system. *Phys. Rev. B* **1989**, *40*, 1795–1805.

39. Jones, A.; Ahmed, W.; Hassan, I.; Rego, C.; Sein, H.; Amar, M.; Jackson, M. The impact of inert gases on the structure, properties and growth of nanocrystalline diamond. *J. Phys.* **2003**, *15*, S2969. [CrossRef]
40. Chen, C.Z.; Qiu, S.H.; Liu, C.Q.; Dan, W.Y.; Li, P.; Ying, Y.C.; Lin, X. Low temperature fast growth of nanocrystalline silicon films by RF-PECVD from SiH_4/H_2 gases: Microstructural characterization. *J. Phys. D Appl. Phys.* **2008**, *41*, 195413. [CrossRef]
41. Lin, K.X.; Lin, X.Y.; Yu, Y.P.; Wang, H.; Chen, J.Y. Measurements in silane radio frequency glow discharges using a tuned and heated Langmuir probe. *J. Appl. Phys.* **1993**, *74*, 4899. [CrossRef]
42. Kondo, M.; Fukawa, M.; Guo, L.; Matsuda, A. High rate growth of microcrystalline silicon at low temperatures. *J. Non-Cryst. Solids* **2000**, *266–269*, 84–89. [CrossRef]
43. Das, C.; Ray, S. Power density in RF-PECVD: A factor for deposition of amorphous silicon thin films and successive solid phase crystallization. *J. Phys. D Appl. Phys.* **2002**, *35*, 2211. [CrossRef]
44. Kirner, S.; Gabriel, O.; Stannowski, B.; Rech, B.; Schlatmann, R. The growth of microcrystalline silicon oxide thin films studied by in situ plasma diagnostics. *Appl. Phys. Lett.* **2013**, *102*, 051906. [CrossRef]
45. Strahm, B.; Howling, A.A.; Sansonnens, L.; Hollenstein, C. Optimization of the microcrystalline silicon deposition efficiency. *J. Vac. Sci. Technol. A* **2007**, *25*, 1198. [CrossRef]

Article

Aluminum Coated Micro Glass Spheres to Increase the Infrared Reflectance

Laura Schwinger [1],*, Sebastian Lehmann [1], Lukas Zielbauer [1], Benedikt Scharfe [2] and Thorsten Gerdes [1]

[1]   Ceramic Materials Engineering-Keylab Glastechnology, University of Bayreuth, 95440 Bayreuth, Germany; Sebastian.Lehmann@uni-bayreuth.de (S.L.); Lukas.Zielbauer@uni-bayreuth.de (L.Z.); Thorsten.Gerdes@uni-bayreuth.de (T.G.)

[2]   Deggendorf Institute of Technology, TAZ Spiegelau, 94518 Spiegelau, Germany; Benedikt.Scharfe@th-deg.de

*   Correspondence: Laura.Schwinger@uni-bayreuth.de; Tel.: +49-921-55-7211

Received: 4 February 2019; Accepted: 8 March 2019; Published: 12 March 2019

Abstract: The reflective properties of micro glass spheres (MGS) such as Solid Micro Glass Spheres (SMGS, "glass beads") and Micro Hollow Glass Spheres (MHGS, "glass bubbles") are utilized in various applications, for example, as retro-reflector for traffic road stripe paints or facade paints. The reflection behavior of the spheres can be further adapted by coating the surfaces of the spheres, e.g., by titanium dioxide or a metallic coating. Such coated spheres can be employed as, e.g., mid infrared (MIR)-reflective additives in wall paints to increase the thermal comfort in rooms. As a result, the demand of heating energy can be reduced. In this paper, the increase of the MIR-reflectance by applying an aluminum coating on MGS is discussed. Aluminum coatings are normally produced via the well-known Physical Vapor Deposition (PVD) or Chemical Vapor Deposition (CVD). In our work, the Liquid Phase Deposition (LPD) method, as a new, non-vacuum method for aluminum coating on spherical spheres, is investigated as an alternative, scalable, and simple coating process. The LPD-coating is characterized by X-ray diffraction (XRD), energy dispersive X-ray spectroscopy (EDX), scanning electron microscopy (SEM), and reflection measurements. The results are compared to a reference PVD-coating. It is shown that both sphere types, SMGS and MHGS, can be homogeneously coated with metallic aluminum using the LPD method but the surface morphology plays an important role concerning the reflection properties. With the SMGS, a smooth surface morphology and a reflectance increase to a value of 30% can be obtained. Due to a structured surface morphology, a reflection of only 5% could be achieved with the MHGS. However, post-treatments showed that a further increase is possible.

Keywords: micro hollow glass spheres (MHGS); solid micro glass spheres (SMGS); liquid phase deposition (LPD); aluminum coating

---

1. Introduction

Micro glass spheres (MGS), such as micro solid glass spheres and micro hollow glass spheres, are utilized in various technical fields due to their excellent physical and chemical properties. Solid Micro Glass Spheres (SMGS) exhibit high strength, as well as smooth surfaces, and are excellently suited as grinding and dispersing balls [1]. They are also utilized as fillers in thermoplastics and thermosets to enhance the physical properties of the matrix, like increasing the Young Modulus and hardness [2–4]. Micro Hollow Glass Spheres (MHGS) are characterized by a low density, as well as by low thermal conductivity. Applied as fillers, e.g., in polymers or building materials, a decrease in weight and minimization of thermal conductivity is possible and could lead to energy savings, especially in the automotive and building industry [5–11].

Possible applications of MGS as additives in the building industry are exterior and interior paints or plasters. MHGS, for example, are processed in facade paints and plasters. SMGS, on the other hand, are used as reflection pigments in marking strips on motorways for improved night visibility in rain and fog as a result of their reflective properties [10,12]. The application as reflection pigment can also be advantageous for wall paints. In the exterior area, reflectance in the visual and near infrared wavelength range (VIS/NIR, 0.4–2.5 µm) plays an important role due to solar radiation, whereas in the interior area, reflection in the mid infrared wavelength range (MIR, 3–50 µm) is important. An increased MIR-reflectance in the range of human black body thermal radiation (~5–30 µm), for example, can increase the thermal comfort in rooms, and thus contribute to saving thermal energy. A beneficial aspect is that the reflection behavior of the spheres can be further modified, e.g., by coating the spheres. For increasing the reflectance behavior in the VIS/NIR range, titanium dioxide coatings are applicable. To improve the reflectance in the MIR-range, metallic coatings, like silver coatings can be applied [13–16].

Due to its properties as a noble metal, silver coatings can be applied rather easily on many materials via electroless deposition [17,18]. Yet there are disadvantages, especially in the application as reflective coating for fillers in paints. First, silver is one of the more expensive precious metals, which would lead to high costs when large amounts of materials are required. Secondly, the high oxidation potential of silver makes silver coatings prone to chemical degradation, which gradually causes an intolerable visible change in the color of the spheres and so of the paint. This also may influence the reflectance [19]. Additionally, silver has a poor recyclability as a color pigment.

In the given case, aluminum is a suitable alternative coating material. It also exhibits a high reflectance in the mid-wave infrared range but is better recyclable and also cheaper than silver. In addition to a low oxidation potential, aluminum films have a good thermal stability and a good adherence to substrates [20]. Thus, none or less influence on the reflection behavior by degradation of aluminum is expected. However, the coating process with aluminum is much more complex than the process with silver. Because of its properties as not noble metal, aluminum cannot simply be deposited by means of electroless deposition. Instead, thin films of aluminum on surfaces are mainly applied via Physical Vapor Deposition (PVD) or Chemical Vapor Deposition (CVD) methods [21–23]. PVD advantageously produces high quality coatings, but is limited in batch size and materials to be coated. For an application as filler on the painting industry, large amounts of coated material are required. However, an upscaling of the PVD process quickly becomes expensive and very complex, not only because of the necessary vacuum process. With the CVD method a wide variety of materials can be deposited with a very high purity and without high vacuum conditions. A disadvantage is that high temperatures (200–1600 °C) are necessary, which results in higher process and energy costs and a limitation of processable substrates. In addition, to generate metallic coatings, metallic-organic precursors are necessary, most of which are highly toxic, explosive, or corrosive, as well as quite costly [23].

A possible alternative to achieve a scalable, more cost-effective process is provided by a chemical coating method, where a precursor is applied in a solvent and deposited on a substrate by means of a decay reaction. Lee at al. [24] have already shown that this Liquid Phase Deposition (LPD) method can be used to produce thin, highly conductive aluminum layers on soda-lime glass and polyethylene terephtalate.

In this paper, the practicability of the LPD method to coat MGS is examined. Applied to hollow glass spheres (MHGS, made of borosilicate glass) and micro solid glass spheres (SMGS, made of soda-lime glass), the achieved coatings are compared to coatings resulting from a PVD process concerning coating quality and homogeneity. Furthermore, the reflection behavior of the LPD-coated spheres is compared with uncoated and PVD-coated spheres in the wavelength range of 7.5 to 18 µm, which in part corresponds to the wavelength range of thermal radiation of a human, as well as to the MIR atmospheric window [25]. The modification of reflection as well as the influence of the different coating techniques and layer qualities on the reflective properties are discussed.

2. Experimental

2.1. Materials

2.1.1. Micro Glass Spheres

Three types of spheres were investigated: Micro Hollow Glass Spheres (MHGS) iM16K-ZF and S38HS from 3M/Dyneon Germany and Micro Solid Glass Spheres (SMGS) Type S from Sigmund Lindner GmbH (SiLi). Relevant properties of these glass spheres are listed in Table 1. The MHGS possess low density and high compressive strength, which can guarantee processability and applicability, e.g., as light filler in wall paints. MHGS type S38HS bear an additional anti-agglomeration agent on their surfaces. To ensure a good comparability between spheres of different consistence, hollow, and solid spheres with a similar reference value D 50 (median of particle size, 50 % of the particles are smaller than the declared value) were chosen.

Table 1. Relevant properties of the examined Micro Solid Glass Spheres (SMGS) and Micro Hollow Glass Spheres (MHGS).

	iM16K-ZF *	S38HS	Type S
Supplier	3M/Dyneon		SiLi
Composition	soda-lime-borosilicate glass		soda lime glass
Sphere type	hollow		solid
Density [g/cm^3]	0.46	0.38	2.50
D 50 [μm] **	20	45	40–70
Compressive strength [MPa]	110	38.5	250–300 ***
Anti-agglomeration agent	No	Yes	unspecified

* laboratory product, not commercially available; ** median of particle size, 50 % of the particles are smaller than the declared value; *** according to the supplier.

2.1.2. PVD-Coated Reference Spheres

PVD-coated spheres are used as a reference material with regard to layer morphology and reflectance. As listed in Table 2, batches of all types of the investigated spheres were coated with different layer thicknesses of aluminum at the University of Vienna. For this purpose, a PVD process especially designed for spherical particles was utilized [21,22]. During the coating process, a strong agglomeration behavior especially for sphere type iM16K-ZF, without anti-agglomeration agent on the surface, was observed. However, these agglomerates could easily be broken up by vibratory plate and sieving without damaging the MHGS.

Table 2. Coating parameters of the aluminum coated glass spheres via Physical Vapor Deposition (PVD) at the University of Vienna.

	iM16K-ZF	S38HS	Type S
Coating volume [l]	1	1	0.1
Coating time [h]	4, 8, 12	8, 16, 21	1, 2, 4
Corresponding layer thickness [nm]	4, 12, 19	12, 29, 34	8, 19, 36

Figure 1 exemplarily shows SEM images of PVD-coated spheres with the highest layer thickness for each type. The entire surface of the spheres is uniformly coated with aluminum and no free spots are present. The surface morphology of the PVD-coated spheres is almost homogeneous with only small imperfections that may be caused either by impurities or by the anti-agglomeration-agent on the surface of the commercial spheres.

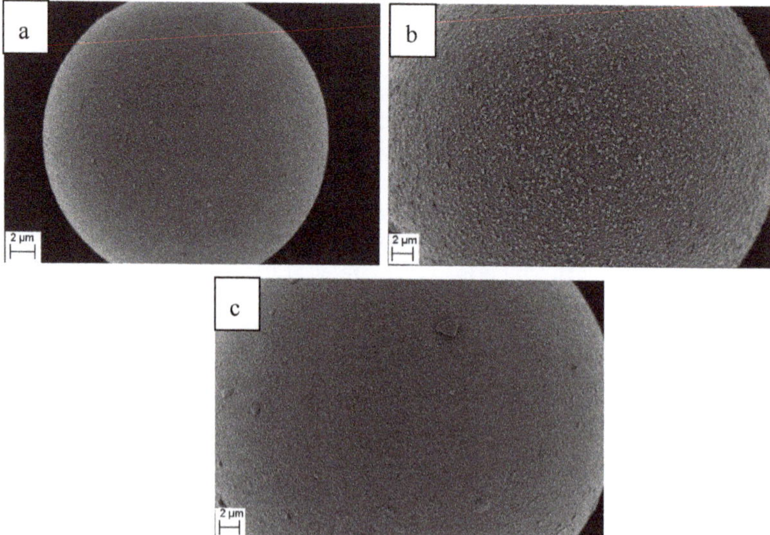

Figure 1. Exemplary SEM-images of the reference micro glass spheres coated via PVD ((**a**): iM16K-ZF: 19 nm; (**b**): S38HS: 34 nm; and (**c**): Type S: 36 nm). An almost homogeneous aluminum coating with some small imperfections on all three types of spheres is achieved on all three types of spheres.

2.2. Liquid Phase Deposition (LPD)

2.2.1. Pre-Conditioning

In contrast to the PVD-coating process, the anti-agglomeration agent as well as possible impurities on the surface of the spheres may act as single seed particles during the LPD-coating process, which favor growth on single spots and thus lead to inhomogeneous layers. To create equal starting conditions for the LPD process and obtain preferably homogeneous aluminum layers, a two-step pre-conditioning of the spheres was performed. First, the MGS were washed three times with ethanol and distilled water, and then dried in a drying cabinet for 24 h at 80 °C to clean the surface of the spheres from impurities. To further support a homogenous coating with aluminum, a calcium silicate nanoparticle layer was created. Following Jin et al. [13], 10 g of the cleaned MGS were then dispersed into a saturated calcium hydroxide solution (Ca(OH)$_2$, Sigma Aldrich) for 4 h at 90 °C. Afterwards, the spheres were filtered and again stored in the drying cabinet for 24 h at 80 °C before coating.

A comparison of untreated MGS to the pre-conditioned MGS is shown in SEM pictures in Figure 2. The untreated surface of the MGS is nearly smooth. The visible small agglomerates (indicated by the white arrows) are either impurities resulting of the manufacturing process, or in the case of the sphere type S38HS, the additional anti-agglomeration agent. After pre-conditioning, calcium silicate nanoparticles are deposited on the surface of the three sphere types. Nevertheless, it appears that more calcium silicate nanoparticles are deposited on the MHGS than on the surface of Type S. This may be due to the different glass types. It was additionally noticed that, for sphere type iM16K-ZF, it agglomerates with sizes of 10–20 spheres that occur after the pre-conditioning. This is explained by the absence of the anti-agglomeration agent. It is presumed that these small clusters are broken up by mechanical means in the later coating process, similar to the reference spheres.

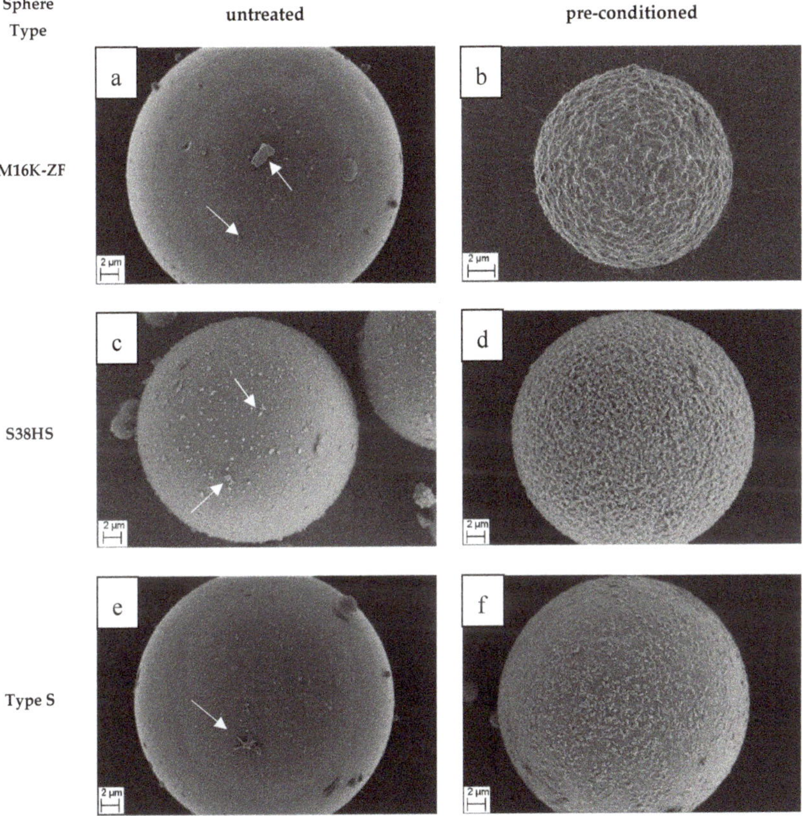

Figure 2. Comparison of untreated Micro Glass Spheres (MGS) and pre-conditioned MGS with Ca(OH)$_2$. The white arrows indicate either impurities resulting from the manufacturing process, or in the case of the sphere type S38HS, the additional anti-agglomeration agent. After pre-conditioning, calcium silicate nanoparticles are deposited on the surface of each sphere type. In contrast to the PVD-coating, the spheres were pre-conditioned to reduce the amount of single seed particles, which would lead to an inhomogeneous aluminum growth. Untreated MGS: (**a**) iM16K-ZF, (**c**) S38HS, (**e**) Type S; pre-conditioned MGS: (**b**) iM16K-ZF, (**d**) S38HS, (**f**) Type S.

2.2.2. LPD of Aluminum

The theoretical reaction step for the LPD of aluminum, as proposed by Lee et al [24] and Brower et al. [26], is shown in Equation (1), the individual reaction steps in Equations (2)–(4).

$$3\, LiAlH_4 + AlCl_3 \xrightarrow{O(C_4H_9)_2} 3LiCl + 6\, H_2 + 4\, Al\,, \tag{1}$$

$$3\, LiAlH_4 + AlCl_3 + 4\, O(C_4H_9)_2 \rightarrow 3\, LiCl + 4\, [OAlH_3(C_4H_9)_2], \tag{2}$$

$$4\, [OAlH_3(C_4H_9)_2] \rightarrow 4\, [OAl(C_4H_9)_2] + 6\, H_2, \tag{3}$$

$$4\, [OAl(C_4H_9)_2] \rightarrow 4\, Al \downarrow + 4\, O(C_4H_9)_2, \tag{4}$$

First the precursor aluminum trihydride dibutyl etherate is produced by reacting 3 M lithium aluminum hydride (LiAlH$_4$, Sigma Aldrich) with 1 M aluminum chloride (AlCl$_3$, Sigma Aldrich) in 2 × 50 mL of anhydrous dibutyl ether (Sigma Aldrich) for 120 min at room temperature (2) [21,22].

The generated precursor is then separated from the byproduct lithium chloride (LiCl) by filtration and subsequently mixed with the Micro Glass Spheres (MGS). Afterwards the precursor is decomposed at a reaction temperature of 130 °C (3,4), where, according to Equation (4), the aluminum supply is related to reaction time as well as the amount of the precursor.

The coating process itself was carried out with two different standard laboratory setups, which also seem appropriate for a technical scaling. In Setup 1, a round flask with magnetic stir bar and reflux condenser under nitrogen atmosphere was employed. Setup 2 comprises a rotating round flask also under a nitrogen atmosphere. For testing and optimizing the coating process of the MGS, several experiments were carried out (Table 3). First, SMGS were coated with different reaction times (120, 240, and 300 min) in both setups to determine a reasonable coating time. Due to the preprocessing step, 10 g of SMGS are employed in each attempt. By transferring this amount of spheres in a graduated cylinder, a bulk volume of 6.5 mL was determined, which has to be taken into account due to the different bulk densities of the investigated sphere types. In addition, experiments with the MHGS were also carried out in both setups.

Table 3. Parameters of the individual coating experiments.

Setup	Experiment	Type of Spheres	Amount of MGS	Reaction Time [min]	Pre-Conditioning	Post-Treatment
1 (with stir bar)	1 A	SMGS	10 g	120 240 300	yes	no
	1B	MHGS	6.5 mL	240	yes	no
2 (rotating flask)	2 A	SMGS	6.5 mL	240	yes	yes
	2 B	MHGS	6.5 mL			yes

2.2.3. Post-Treatment

As for the PVD-coated spheres, agglomerations were also observed in the PVD process. As mentioned before, the agglomerated spheres can be broken up by mechanical means. Therefore, different post-treatments were applied to the LPD-coated spheres to further investigate the disaggregation of clustered spheres. Simultaneously, a possible influence on the surface morphology and the stability of the aluminum coating on the sphere surface can be tested.

Turbula®Mixer:

Agglomerates of the PVD-coated spheres could be broken up by vibrational movement, so the LPD-coated spheres were first treated with a Turbula®Mixer (Willy A. Bachofen AG, Muttenz, Switzerland). Different mixing times (2, 4, 6, 8, 12, and 24 h) were investigated.

Magnetic Stir Bar Treatment:

To further examine the mechanical influence of the magnetic stir bar utilized in Setup 1, the coated spheres were placed in water and stirred for 12 h. Water was chosen as solvent as it allowed the simultaneous investigation of how the coated spheres behave in a slightly alkaline milieu. This behavior is interesting for a later possible use as additive in water-based wall paints.

Furnace Treatment:

In addition to the mechanical influences, a thermal influence on the coated spheres was also investigated.

A possible homogenization or smoothing of the coating layer by sintering or melting is pursued by heating up coated spheres in a furnace to different temperatures with a gradient of 10 °C/min in a forming gas (Ar/H$_2$ 95/5) atmosphere. Two temperatures below (400 and 500 °C) and one temperature above (750 °C) the melting temperature T_m of aluminum (approx. 660 °C) were chosen. The holding time was set to 2.5 h.

2.3. Experimental Characterization Methods

To determine coating quality and reflective properties, each batch of spheres was subjected to a detailed experimental investigation by the following characterization methods.

2.3.1. Qualitative Phase Composition

The coating was analyzed by X-Ray structure analysis (XRD) to determine if pure aluminum was deposited on the surfaces of the spheres. It was carried out with an X'-Pert MPD PW 3040 X-ray diffractometer from Philips (Philips Co., Netherlands). The individual samples were measured using Cu Kα radiation in a range from 10–90° 2Θ. The step speed was 0.03° with a holding time of 30 sec. The obtained spectra were analyzed with the software X' Pert High Score Plus 4.1 (Malvern Panalytical, Malvern, UK).

To analyze the elemental phase composition of the coating, an Energy Dispersive X-Ray analysis (EDX) was conducted. The measurements were carried out with a Thermo Fisher Ultra Dry SDD Silicium Drift Detector (Thermo Fisher, Darmstadt, Germany) with an excitation voltage of 20 kV.

2.3.2. Morphology Characterization

The morphology of the coated spheres was observed using a Zeiss LEO 1530 FESEM scanning electron microscope (SEM) (Zeiss, Oberkochen, Germany) at the Bavarian Polymer Institute (BPI) of the University of Bayreuth. The excitation voltage was set to 3 kV. In advance, the samples were sputtered with a 1.3 nm thick platinum layer by a 208HR sputter coater (Cressington Scientific Instruments, Dortmund, Germany).

2.3.3. Reflection Measurement

MIR-range reflection measurements were performed with a FTIR instrument (Vertex70 from Bruker Optic GmbH, Ettlingen, Germany) and an integrating sphere (U-Cricket TM-BR4 from Harrick Scientific Products, Darmstadt, Germany) under a nitrogen atmosphere against a gold standard. The measuring range of the Vertex 70 is approx. 600 to 7000 cm^{-1}, which corresponds to a wavelength range of approx. 1.5 to 18 µm. The samples were applied as a monolayer to an adhesive strip and then placed on the opening of the integrating sphere. The reproducibility was confirmed by repeated preparation and measurements of the powder samples. An error of less than 0.01% was determined.

3. Results and Discussion

3.1. Morphological Examination

3.1.1. Setup 1—Round Flask with Magnetic Stir Bar

Experiment 1 was carried out in a round flask with a magnetic stir bar and reflux condenser under a nitrogen atmosphere. The color change from white to gray after each coating test qualitatively demonstrates the metallization of the spheres (Figure 3)

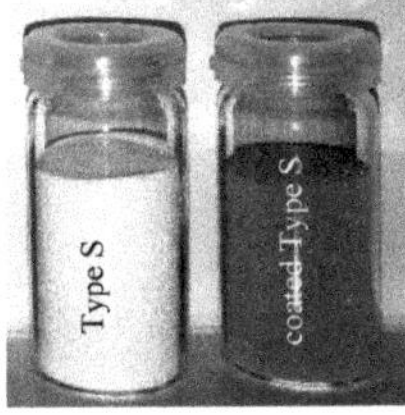

Figure 3. Coating of Type S spheres after a coating time of 240 min: The original spheres have a white color, the Al-coated spheres show a gray coloration.

In Experiment 1 A coating tests with 10 g of spheres of Type S and coating times of 120, 240, and 300 min were performed. As indicated in Figure 4, a coating is achieved for all investigated times.

However, an increased coating time leads to smoother surfaces. The interaction of the friction of the magnetic stir bar, which is used for mixing of the coating solution during the coating process, and the necessary reaction time for a complete decomposition are identified as possible reasons.

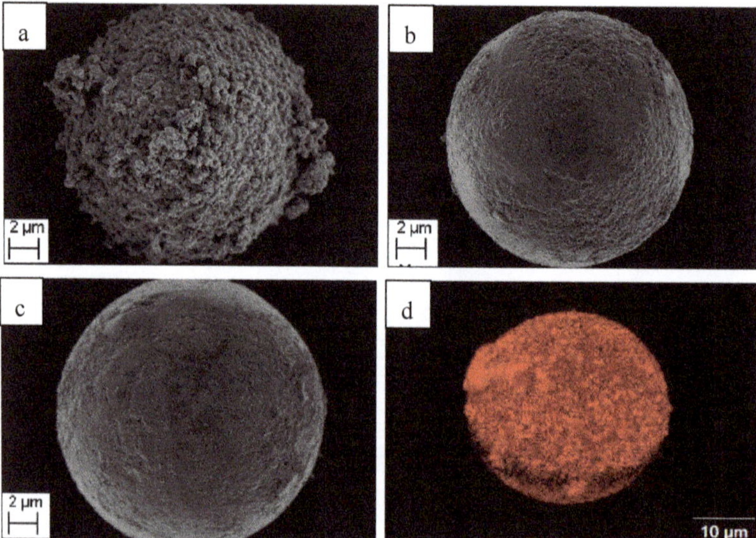

Figure 4. SEM images of Type S spheres coated via LPD-process (Experiment 1, 10 g) at different coating times: (**a**): 120 min; (**b**): 240 min; and (**c**): 300 min. As the coating time increase, the aluminum coating morphology becomes smoother due to the reaction process and the influence of the magnetic stir bar; (**d**) EDX-Mapping of a coated sphere Type S.

After 120 min, the reaction does not seem to be completed, and therefore, additional aluminum was deposited on the surface of the spheres despite the friction of the stir bar. Between 120 and 240 min, however, the reaction was completed and the friction of the stir bar smoothed the surface. It also becomes clear that the aluminum must be firmly fixed to the sphere surface, since no abrasion of the coating takes place. This will be examined more closely in the section of the post-treatment (3.1.3). As only little differences are obtained between coating times of 240 and 300 min, the coating time of 240 min was chosen for further experiments.

Compared to the PVD-coated spheres, coating times of 240 and 300 min results in a homogeneous coating and smooth surface morphology. The uniform application of the aluminum coating is also confirmed by the EDX mapping. The results of X-Ray diffraction analysis are plotted in Figure 5. The uncoated spheres show an amorphous halo in the range of 20–40 Θ, which is characteristic for amorphous materials like glass. Besides the halo, the aluminum coated samples deliver characteristic peaks of metallic aluminum at 38° (111), 45° (200), 65° (220), 78° (211), and 82°(222). Thus, it can be verified that pure metallic aluminum is deposited on the sphere surfaces with the liquid phase deposition (LPD) method.

Exemplary EDX-Mapping-Measurements (d) of a coated sphere of Type S (Experiment 1) shows a uniform aluminum coating (red) on the surface of the spheres after a coating time of 240 min.

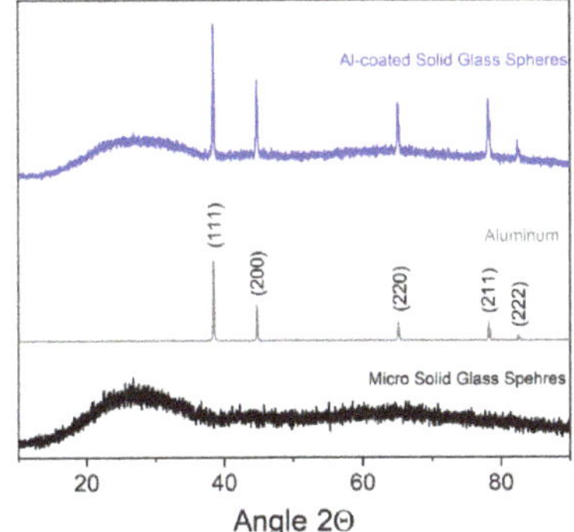

Figure 5. Qualitative comparison of X-Ray Diffraction of uncoated and LPD-coated Type S spheres. Besides the characteristic halo of amorphous materials, all characteristic peaks of metallic aluminum are detected on the coated spheres.

While for the spheres of Type S a rather homogenous and smooth coating was possible, attempts to coat the MHGS spheres type S38HS with Setup 1 was not successful. Contrary to the SMGS, the MHGS are floating on top of the reaction solution due to their significant lower density. To enforce a complete mixing, the speed of the magnetic stir bar had to be increased. The higher shear forces caused an increased breakage of the spheres, especially for type S38HS. This is justified by the low compressive strength of this sphere type (Table 1). Due to the higher compressive strength, type iM16K-ZF did not show an increased sphere breakage under these conditions, but a strong agglomeration and a very inhomogeneous coating were observed.

3.1.2. Setup 2—Rotating Flask

To eliminate the damage of the spheres by the magnetic stir bar and the influence of floating, especially for the MHGS, Experiment 2 A and 2 B were carried out in a rotating flask under nitrogen atmosphere. Again, 10 g or 6.5 mL of spheres of Type S were coated to obtain a comparison to the results attained with round flask and magnetic stir bar. Figure 6 left exemplarily shows a SEM image of a coated sphere of Type S. Compared to Setup 1 partly bigger agglomerations stick on the surface (indicated by white arrows). In addition, clumping of the spheres take place. This is explained with the absence of the magnetic stir bar, which smooths the surfaces of the spheres by friction and probably is breaking up agglomerates. This will be examined in more detail in Chapter 3.1.3. Nevertheless, also uniform coated spheres could be obtained with Setup 2, which is confirmed by the EDX mapping (Figure 6b).

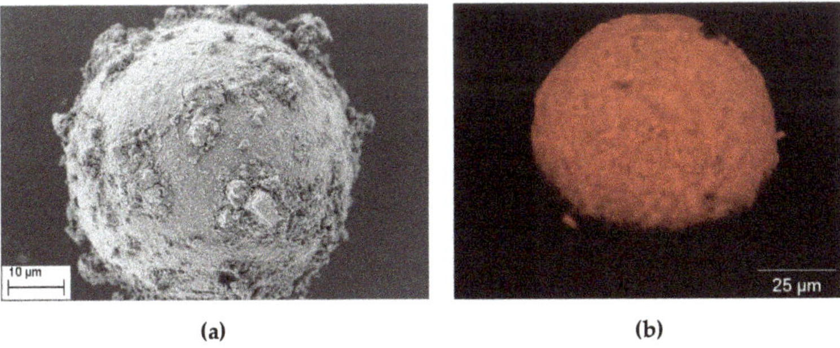

(a)   (b)

Figure 6. (**a**): SEM image of Type S coated via LPD-process in a rotating flask (Experiment 2 A). A complete aluminum coating with bigger agglomerates was achieved; and (**b**): EDX-Mapping of a coated sphere Type S (Experiment 2 A). A homogenous aluminum coating (red) could be proven on the surface of the sphere.

In Experiment 2 B a defined volume of 6.5 mL of each type of MHGS was coated. The color change of the spheres from white to gray after each coating test once again also indicates a successful metallization of the spheres (Figure 7).

Figure 7. Qualitative verification of the coating of the MHGS with Setup 2: The original spheres have a white color, the coated spheres show a gray coloration.

SEM images and EDX mappings in Figure 8 confirm that a homogenous coating on both types of MHGS can be reached in Experiment 2B. No ball breakage occurs. The XRD results again indicate a metallic aluminum coating (Figure 9).

In comparison with the reference spheres and the coated spheres of Type S of Experiment 1 A, however, a more structured morphology results. This is explained by the reduced friction in the rotary flask compared to the stirring, which led to a smoother surface structure in Experiment 1. In addition, both types of MHGS now show a partly agglomeration (10–20 spheres). This, nonetheless, is not a problematic result as the post-treatment is intended to break up these agglomerates.

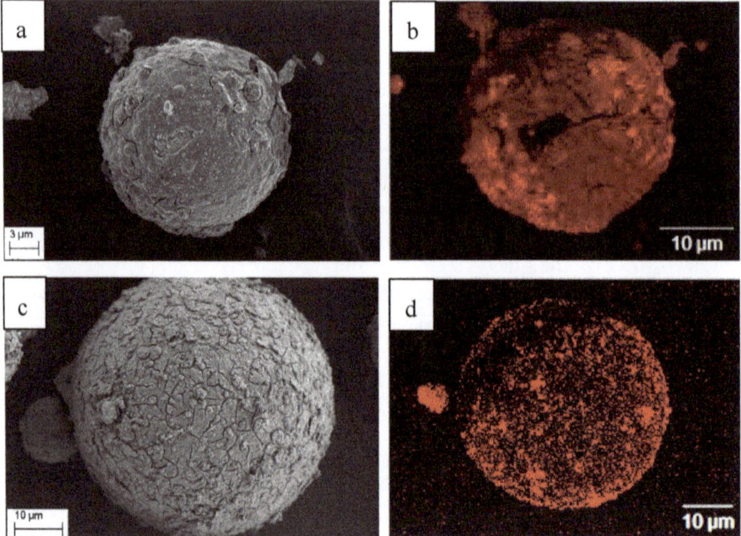

Figure 8. SEM images and EDX-Measurements of MHGS ((**a,b**): iM16K-ZF and (**c,d**): S38HS) coated via LPD-process in a rotating flask (defined volume: 6.5 mL, Experiment 2 B). In comparison to the reference spheres, a more structured aluminum coating could be obtained on both types of spheres.

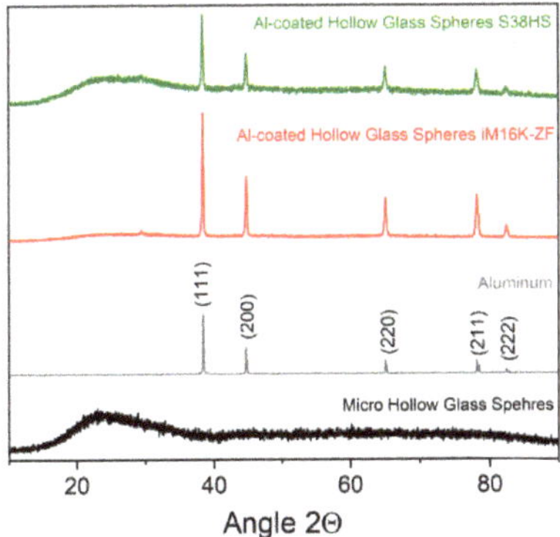

Figure 9. Qualitative comparison of X-Ray Diffraction of uncoated and LPD-coated MHGS. Besides the characteristic halo of amorphous materials, all characteristic peaks of metallic aluminum are detected on the coated spheres.

3.1.3. Post-Treatment

In the previous chapters, it was shown that, with the LPD-method, it is possible to metallize a rather uniform aluminum layer on micro hollow and micro solid glass spheres. Different results regarding layer morphology were achieved and partly agglomeration was observed. As agglomerates of PVD-coated spheres could be broken up by mechanical means, additional post-treatments were carried out to investigate whether the small agglomerates of the LPD-coated spheres can also be broken up by mechanical means. In addition, the stability of the coating as well as the surface morphology after the treatments were examined. Coated sphere types iM16K-ZF and Type S of Experiment 2 were used for the post-treatments.

As already investigated, friction by the magnetic stir bar effects a smoothing of the aluminum coating of the SMGS, whereas MHGS show increased ball breakage. Friction in the rotating flask caused no breakage of the MHGS but is not sufficient to smooth the aluminum coating. Therefore, the rotating friction, simulated in a Turbula®Mixer, is increased by raising the time in order to achieve a smoothing effect. In addition, agglomerated spheres could be broken up. Hence, the coated spheres were first treated in a Turbula®Mixer for different times (2, 4, 6, 8, 12, and 24 h).

Figure 10 exemplarily shows SEM-images of spheres of Type S treated for 24 h. The agglomerates present in the untreated sample are almost completely broken up by the treatment. This can also be seen with sphere type iM16K-ZF. Thus, with Setup 1 and Setup 2, single coated spheres can be obtained. This is advantageous, because single coated MGS can be processed more easily. Further, the overall surface is increased, which is positive for the reflective behavior. In addition, there is also a small change in the surface morphology. The previously structured surface can be smoothed slightly, which in turn again can have a positive effect on the reflective behavior. The aluminum coating is firmly bonded to the surface of both spheres, and no break off of the coating results by friction. This can have a positive influence on the potential use in wall paints, where fillers are also included in paints by mixing.

Figure 10. SEM images of the untreated ((**a,c**) and post-treated (Turbula®Mixer; (**b,d**)) SMGS Type S (Experiment 2 A). Breaking off of the agglomerates and no change in surface morphology is evident.

A second post-treatment method was carried out to investigate the influence of the magnetic stir bar in Setup 1. It was already shown that the magnetic stir bar breaks up the agglomerates during the coating process and smooths the surface of the spheres. It was tested whether this is also possible afterwards for coated spheres of Setup 2. Therefore, coated spheres of Type S (Experiment 2A) were transferred into distilled water and stirred for 12 h. As explained above, distilled water was used to equally test the influence of a slightly basic milieu. No color change of the post-treated spheres occurs, which means that no influence on the aluminum layer results from the slightly basic milieu. Again, the agglomerates could be broken up by the post-treatment with magnetic stir bar. Additionally, it can be recognized in Figure 11 that there is a partially significant change of the surface structure. On some spheres, a very smooth surface, like in Experiment 1, results. The fact that it only occurs partially may be due to different sphere-solvent ratios (2 g/100 mL) compared to the coating test (10 g/100 ml). Thus, it is possible that not all spheres get the same level of friction. This treatment, however, could only be carried out with the spheres of Type S, as the problem of floatation occurred with the sphere type iM16K-ZF.

Considering the stability of the coating, no flaking or destruction of the aluminum layer took place even after 12 h. Thus, the aluminum coating sticks firmly on the surface of the spheres of Type S and can also be smoothed by a post-treatment with a magnetic stir bar.

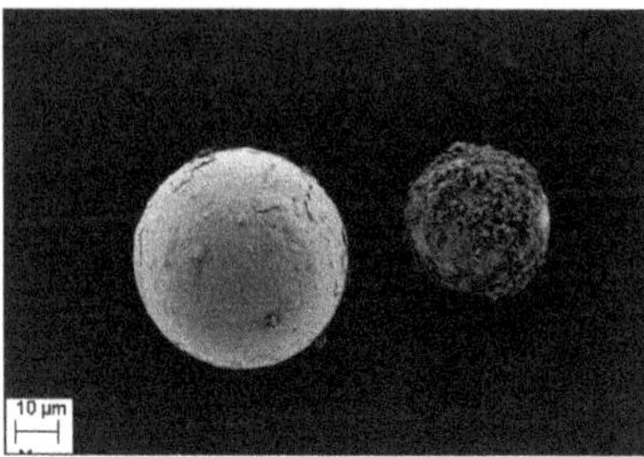

Figure 11. SEM images of the post-treated (with magnetic stir bar, after 12 h) SMGS Type S (Experiment 2 A). Partially, a significant change in morphology takes place.

Since the agglomerates can be broken up and the surface morphology can be changed by friction, a third treatment was carried out to investigate whether a thermal treatment has an influence on the morphology of the coating, too. As already shown in Experiment 2 the coated spheres showed a more structured surface morphology. The furnace treatment is intended to show whether smoothing the coating is possible by sintering or melting. Therefore, the post-treatment method consists of heating up the coated spheres to different temperatures (400, 500 and 750 °C, 10 °C/min, no holding time, Ar/H_2 95/5 atmosphere) in a furnace.

Exemplarily SEM images of coated Type S spheres are shown in Figure 12. For furnace temperatures of 400 and 500 °C no significant change of the aluminum layer occurs. Agglomerates of single spheres were formed, indicating that a sintering of the layer took place. The sticking together could possibly be prevented by a rotating chamber. However, the temperature is not sufficient to completely smooth the aluminum layer on the surface of the spheres. For a furnace temperature of 750 °C, in contrast, a noticeable change in the aluminum layer can be detected. Smaller crystal-like and worm-like structures were formed, which makes the sphere surface much more inhomogeneous than it was before. This structural change is caused by the recrystallization of aluminum, as was also described from K. D. Vanyukhin et al. [27]. A melting and thus a smoothing of the coating does not take place.

Figure 12. SEM images of the post-treated (furnace) SMGS Type S (Experiment 2 A). No significant change in surface morphology at a temperature of 400 °C (**a**) and 500 °C (**b**) is apparent. For a temperature of 750 °C (**c,d**) a more inhomogeneous surface occurs due to the recrystallization of aluminum.

Since neither friction nor heat causes a break off or flaking of the aluminum layer, it can be assumed that a stable aluminum coating results from the LPD method. A positive effect of the post-treatment with the Turbula®Mixer and the magnetic stir bar, that agglomerates can be broken, which makes the spheres more easily processable and increases in reflective surface. Additionally, a partially significantly change in surface morphology occurred for the coated SMGS. Thus, through sufficient friction, a smooth coating morphology, similar to that of the reference spheres, can be achieved on solid glass spheres.

3.2. Reflection Measurements

In Chapter 3.1, it was shown that it is possible to get a rather homogeneous aluminum coating on micro hollow and solid glass spheres with the LPD method. Since the aim of the work is not only to investigate the possibility to metallize the MGS, but also to explore the increase of the reflection in the MIR wavelength range, reflection measurements were carried out in addition to the optical and morphological evaluation using SEM images and as well as XRD and EDX. The question whether the reflection can be increased due to the properties of aluminum in contrast to the uncoated MGS was investigated. Further, the reflectance of the LPD-coated spheres and PVD-coated spheres was compared to examine the influence of different surface morphologies.

As an adhesive strip is utilized to apply the powdery samples to the opening of the integrating sphere of the measurement device, a possible influence of the strip was first examined. As can be seen in Figure 13 (black curve), the adhesive strip has a nearly 0% reflectance in the chosen measurement range between 7.5 and 18 µm and thus can be neglected. The samples were sieved beforehand to ensure that only coated spheres and no Al dust were measured. Furthermore, the influence of the pre-conditioning with $Ca(OH)_2$ was investigated by comparing the reflectance of pre-conditioned MHGS and SMGS with untreated spheres in the same range (Figure 13). Below 10 µm, soda-lime and borosilicate glasses have absorptions bands, above reflectance values between 3% and 10% arise.

The influence of the pre-conditioning can be neglected for the sphere type Type S, since the pre-conditioned micro spheres exhibit nearly the same reflectance in the wavelength range of interest as the original spheres (Figure 13, blue curves). For the sphere types iM16K-ZF (Figure 13, red curves) and S38HS (Figure 13, green curves), a slight difference in the reflection can be observed between 9 and 11 µm and for wavelengths greater than 17 µm. The latter one is explained by the resolution limit of the device, the differences in the smaller wavelength range may be due to an inhomogeneous coating.

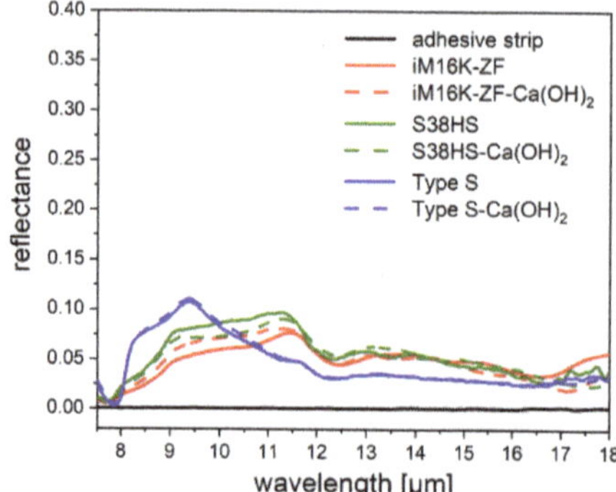

Figure 13. Reflectance behavior of the adhesive strip, the original spheres and the pre-conditioned spheres in the wavelength range of 1.6 to 25 µm. Less or no influence of the adhesive strip and pre-conditioning in the wavelength range of interest (10 to 18 µm) is assumed.

3.2.1. PVD-Coated Spheres

First, the reflectance values of the PVD-coated reference spheres were determined with the described setup and a correlation between different layer thicknesses and reflection was evaluated. At lower layer thicknesses, the curve still resembles the reflection curve progression of uncoated spheres. With increasing layer thickness, almost constant reflection values emerge, which is typical for aluminum (Figure 14a–c). Thus, the almost constant curve-shape indicates a sufficient deposition of aluminum on the surface of the spheres, as no features of glass reflection are observed any more [23]. The decrease in reflectance observed for wavelength values greater than 17 µm is again related to the resolution limit of the measurement device.

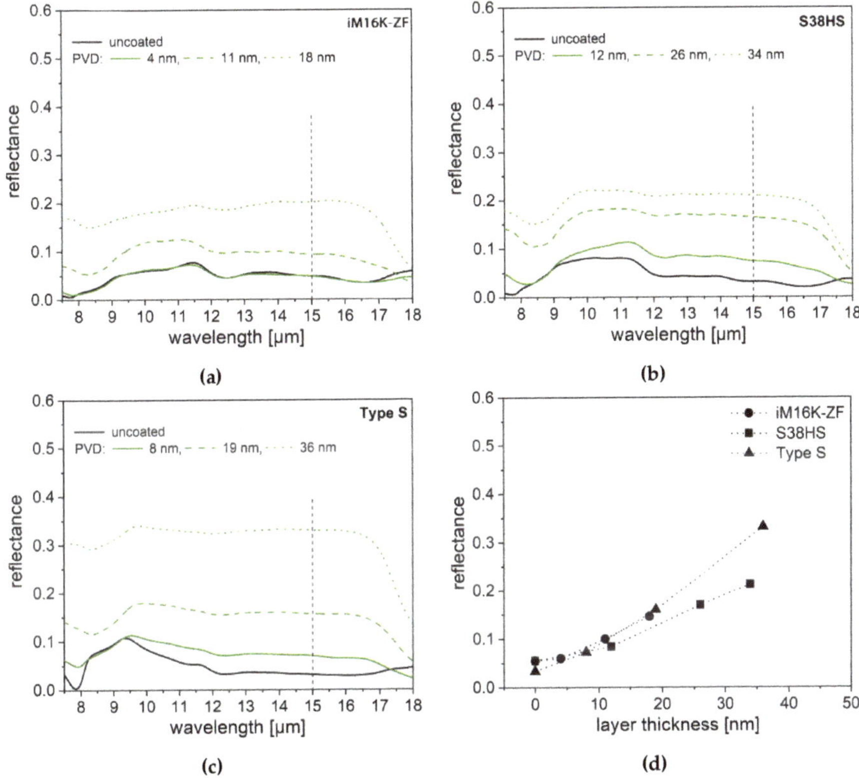

Figure 14. Results of the reflection measurements of the PVD-coated MGS (reference spheres, (**a**): iM16K-ZF, (**b**): S38HS, (**c**): Type S) as well as the reflection at a wavelength of 15 µm of each sphere types plotted over the layer thickness (**d**). An increased reflection with increasing layer thickness can be observed.

It becomes apparent that, with the increasing layer thickness, an increase in reflection also takes place for all three sphere types. To show this more clearly, in Figure 14d the reflectance values of each sphere type at a wavelength of 15 µm is plotted against the layer thickness. A similar behavior was observed by Lugolole et al. [20]. Interestingly, a similar layer thicknesses of approx. 36 nm lead to different reflectance values. A higher reflection degree of the SMGS in comparison to the MHGS is achieved.

3.2.2. LPD-Coated Spheres—Setup 1

In Figure 4, it was already shown that a very homogeneous, smooth aluminum coating can be achieved with Setup 1 and a coating time of 240 min for the spheres of Type S. The results of the reflection measurements of these LPD-coated spheres and the comparison with the PVD-coated spheres are represented in Figure 15.

Considering the reflectance measurements, it can be seen that a significant increase in reflectance is observed as a function of the coating time from approx. 5% to 35%. However, after a coating time of 300 min, no more increase of the reflectance occurs in comparison to a coating time of 240 min. It was already examined in Chapter 3.1.1, how this is in the line with no change in surface morphology of the two samples, as the available aluminum has already been completely deposited.

Compared to the reflectance of the reference spheres (layer thickness 36 nm) similar values are achieved for the LPD-coated spheres. Thus, using the LPD method for Type S, both a similar surface morphology and reflectance in comparison to the PVD-coated Type S can be obtained.

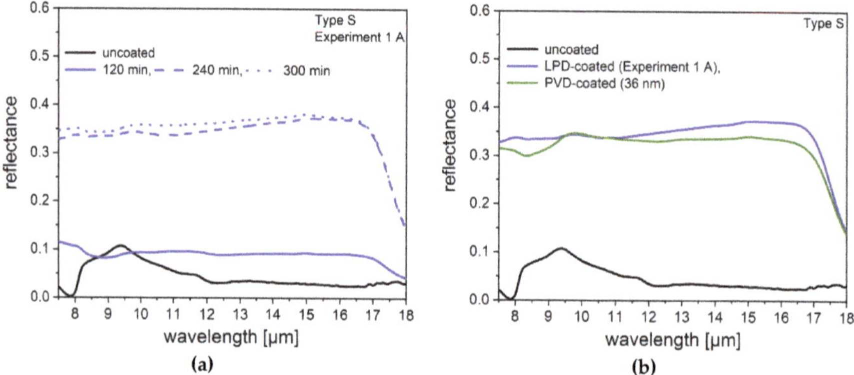

Figure 15. (a): Results of the reflection measurements of the LPD-coated Type S spheres (Experiment 1) with different coating times. Reflection increases with increasing coating time. **(b)**: Comparison of the reflectance of LPD-coated (Experiment 1 A, 240 min) and PVD-coated (36 nm) spheres of Type S.

3.2.3. LPD-Coated Spheres—Setup 2

The results of the reflectance measurements of the coated spheres from Experiment 2 are shown in Figure 16. It was already established by SEM images Figures 6 and 8 that all three types of spheres could be coated with aluminum, but in comparison to the reference spheres, a partly rougher surface morphology occurs. Despite the coated layer, for both types of spheres, no or less reflectance increase can be observed. This behavior was already established by Lugolole et al. [20], where the lowest reflectance was achieved with the sample, which has a rough surface. Das et al. [28], Moushumy et al [29], and Sharma et al [30] also show numerical simulations of different nano-structured gratings on GaAs substrates, that the surface structure has a significant influence on the reflection behavior. With a flat substrate, a reflection of about 28% can be achieved, whereas a structured surface shows only a reflection of about 2% [28].

Nevertheless, the reflectance curves of the coated spheres show an almost constant progress over the wavelength range and compensate at least partially the reflectance curve of the uncoated glass. This indicates that at least a sufficient amount of aluminum is coated on the spheres. Therefore, the uneven layer structure of the aluminum coating must be the determining factor.

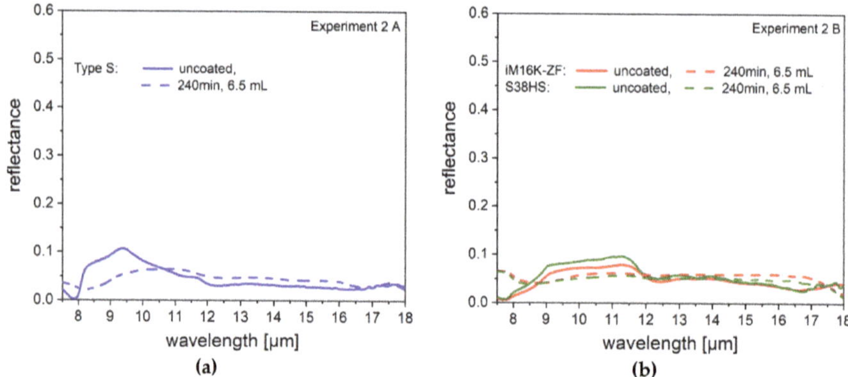

Figure 16. (a): Results of the reflection measurements of the LPD-coated SMGS (Experiment 2 A) with an amount of 10 g. No increase in reflectance occurs. **(b)**: Results of the reflection measurements of the LPD-coated MHGS (Experiment 2 B) with a volume of 6.5 mL. Further, no increase in reflection occurs.

3.2.4. Post-Treated Spheres

As can be seen in Figure 17, the reflectance can generally be increased by all three post-treatments. They led to a partial change in surface morphology towards smoother surfaces. Only in the case of the furnace treatment over the melting temperature of aluminum, a surface deterioration was observed. Here the dependence of reflectivity surface can again be clearly determined. The very inhomogeneous surface structure (Figure 12a,b) which is reflected in a clear decrease in reflectance to almost 1%. On the contrary, the furnace treatment below the melting point is associated with a reflection increase of approx. 3% (Type S) and approx. 2% (iM16K-ZF), despite no change in the SEM images was observed. The partially smoothed surfaces after magnetic stir bar treatment lead to a slightly better increase of 5% (Figure 17c). Besides surface smoothening, agglomerates of spheres could be broken up by mixing. This also should increase the reflection of the spheres due to the enlargement of the total surface. After 24 h in the Turbula®Mixer, an actual increase in reflection of approx. 3% occurred for the spheres of Type S (Figure 17d). For type iM16K-ZF a similar increase of 2% could be achieved. Thus, the dissembling of the agglomerates not only facilitates the processability, but also leads to an increase in reflectance.

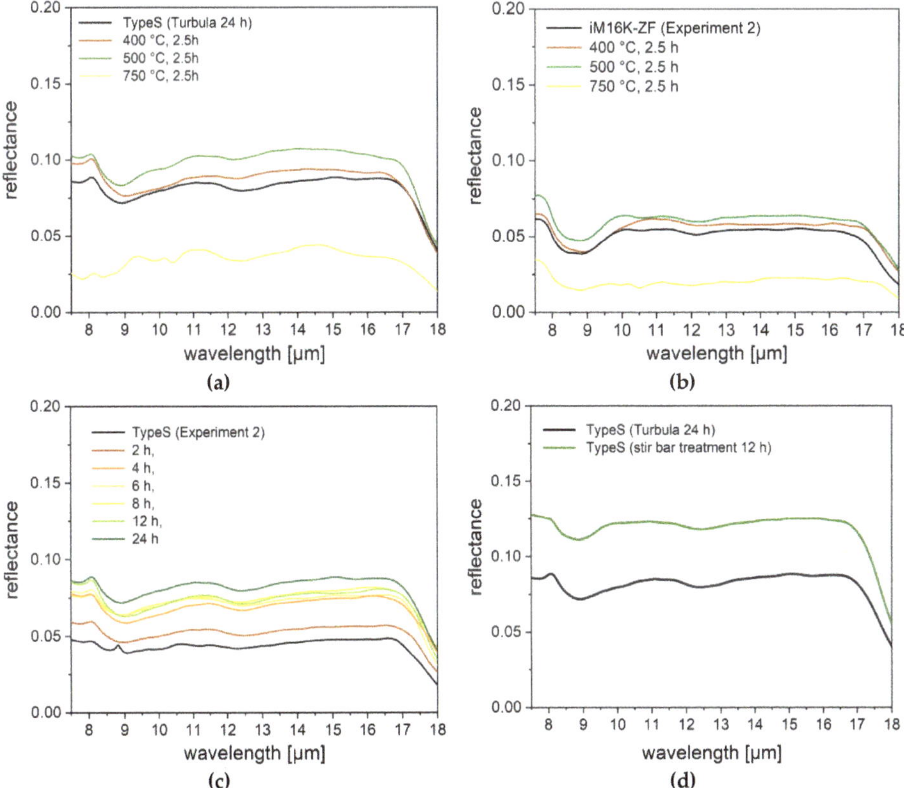

Figure 17. Measured reflectance of the post treated MGS from Experiment 2 A and 2 B, (**a,b**): Furnace treatment; (**c**): Turbula®Mixer, and (**d**): Stir bar treatment, an increase in reflection can be determined for all three post-treatments.

In conclusion, the reflection measurements show that a change in morphology and thus an increase in reflection can be achieved through post-treatments. Despite the post-treatments, however, the result from Experiment 1 could not be tied up. Therefore, further works will be required to find a post-treatment that adapts the surface morphology for the coated MGS from Setup 2.

4. Summary

In this paper, the Liquid Phase Deposition (LPD) method was examined as an additional coating method to the Physical Vapour Deposition (PVD) and the Chemical Vapor Deposition (CVD) method for metallization of micro glass spheres. Two different micro hollow glass sphere types (MHGS, iM16K-ZF and S38HS) and one micro solid glass sphere type (SMGS, Type S) were coated with aluminum by the hydridolyse of $AlCl_3$. Measurements of X-Ray Diffraction (XRD), Energy Dispersive X-Ray Spectroscopy (EDX) and Scanning Electron Microscope (SEM) show, that both sphere types (MHGS and SMGS) can be coated with metallic aluminum using the Liquid Phase Deposition (LPD) method. In comparison to reference spheres, coated via the Physical Vapor Deposition (PVD) method, a similar surface can be achieved for the SMGS, whereas a more structured morphology occurred for the MHGS. Post-treatments with Turbula®Mixer, magnetic stir bar and furnace exhibit that the aluminum layer is stable and firmly bonded to the sphere surface. In addition, the post-treatments were able to break up agglomerates, as well as slightly smooth the surface morphology.

The influence of the coating and the post-treatments on the reflective properties in the MIR-range of the spheres was examined by additional reflection measurements in a wavelength range of 7.5 to 18 μm. It was found that in addition to a sufficient amount of aluminum, surface properties play an important role. With a rather rough surface morphology, only an increase in reflectance of approx. 5% was observed for SMGS. In contrast, a very homogenous smooth surface resulted in an increase in reflectance of approx. 30 % compared to the uncoated spheres, which corresponds to the reflectance of the PVD-coated spheres. However, this cannot be obtained with the MHGS. However, the post-treatments showed that an adaptation of the surface morphology is possible, leading to double the reflectance compared to the untreated coated spheres.

The paper showed that the LPD method, as a non-vacuum process, is suitable for an aluminum coating of spherical particles. It was even possible to achieve coating results comparable to the PVD process, which was shown by the similar reflection values of the LPD-coated spheres and PVD-coated spheres. Besides reflection, aluminum coated (hollow) glass spheres can also be interesting for further application, e.g., the modification of thermal conductivity or dielectric properties, where, unlike in our case, the surface condition will have less influence.

Author Contributions: L.S. performed the experiments and wrote the paper. L.Z. and B.S. helped with the experiments. S.L. und T.G. provided guidance and helped in manuscript preparation.

Funding: Financial support by the BMBF project EcoSphere (BMBF: 13N13188) is gratefully acknowledged.

Acknowledgments: The presented work is dedicated to Monika Willert-Porada, who passed away on 11 December 2016. She initiated the above-mentioned project. In addition, the authors would like to thank Angelika Kreis and Ingrid Otto for laboratory support. The authors also want to thank Andreas Eder from the University of Vienna for producing the PVD-coated spheres. This publication was funded by the German Research Foundation (DFG) and the University of Bayreuth in the funding program Open Access Publishing.

Conflicts of Interest: The authors declare no conflict of interest.

References

1. Wypch, G. *Handbook of Fillers*; Chem Tec Publishing: Toronto, ON, Canada, 2016; p. 4. ISBN 9781895198911.
2. Kulshreshtha, A.K.; Vasile, C. *Handbook of Polymer Blends and Composites*; Rapra Tehnology Ltd.: Shrewsbury, UK, 2003; p. 4B. ISBN 9781859573044.
3. He, D.; Jiang, B. The elastic modulus of filled polymer composites. *J. Appl. Polym. Sci.* **1993**, *49*, 617. [CrossRef]
4. Lee, J.; Yee, A.F. Fracture of glass bead/epoxy composites: On micro-mechanical deformations. *Polymer* **2000**, *41*, 8363–8373. [CrossRef]

5. Yung, K.C.; Zhu, B.L.; Yue, T.M.; Xie, C.S. Preparation and properties of hollow glass microsphere-filled epoxy-matrix composites. *Compos. Sci. Technolol.* **2009**, *69*, 260–264. [CrossRef]

6. Xu, N.; Dai, J.; Zhu, Z.; Huang, X.; Wu, P. Synthesis and characterization of hollow glass–ceramics microspheres. *Ceram. Int.* **2011**, *37*, 2663–2667. [CrossRef]

7. Li, B.; Yuan, J.; An, Z.; Zhang, J. Effect of microstructure and physical parameters of hollow glass microsphere on insulation performance. *Mater. Lett.* **2011**, *65*, 1992–1994. [CrossRef]

8. Drobny, J.G. *Handbook of Thermoplastic Elastomers*; Elsevier: Amsterdam, The Netherlands, 2014; Volume 1, p. 25. ISBN 9780323221368.

9. Weatherhead, R.G. *FRP Technology Fibre Reinforced Resin Systems*; Springer: Berlin/Heidelberg, Germany, 2012; Volume 1, p. 381. [CrossRef]

10. Budov, V.V. Hollow Glass Microspheres. Use, Properties and Technology (Review). *Glass Ceram.* **1994**, *51*, 230–235. [CrossRef]

11. Lehmann, S.; Schwinger, L.; Scharfe, B.; Gerdes, T.; Erhardt, M.; Riechert, C.; Fischer, H.-B.; Schmidt-Rodenkirchen, A.; Scharfe, F.; Wolff, F. Mikro-Hohlglaskugeln als Basis Energieeffizienter Dämmung von Gebäuden. In Proceedings of the HighTechMatbau Conference, Berlin, Germany, 4–5 December 2018; p. 21.

12. Kemsley, J.N. What's that stuff? Road Markings-Pigments, polymers, and reflective spheres help keep you safe on the road. *Chem. Eng. News* **2010**, *88*(36), 67.

13. Jin, H.; Xu, D.; Li, J.; Wang, K.; Bi, Z.; Xu, G.; Xu, X. Highly Solar-Reflective Litchis-Like Core-Shell HGM/TiO$_2$ Microspheres Synthesized by Controllable Heterogeneous Precipitation Method. *NANO Brief Rep. Rev.* **2017**, *12*, 1750080. [CrossRef]

14. Willert-Porada, M. (Ed.) *Second Progress Report Forglas*; Subproject I.2; Universität Bayreuth: Bayreuth, Germany, 2011; pp. 51–74.

15. Mohelnikova, J. Materials for reflective coatings of window glass applications. *Constr. Build. Mater.* **2009**, *23*, 1993–1998. [CrossRef]

16. Cozza, E.S.; Alloisio, M.; Comite, A.; di Tanna, G.; Vicini, S. NIR-reflecting properties of new paints for energy-efficient buildings. *Sol. Energy* **2015**, *116*, 108–116. [CrossRef]

17. Zhang, H. Silver plating on hollow glass microsphere and coating finishing of PET/cotton fabric. *J. Ind. Text.* **2012**, *42*, 283–296. [CrossRef]

18. Xu, X.; Luo, X.; Zhuang, H.; Li, W.; Zhang, B. Electroless silver coating on fine copper powder and its effects on oxidation resistance. *Mater. Lett.* **2003**, *57*, 3987–3991. [CrossRef]

19. Morales, A.; Durán, A. Sol-Gel Protection of Front Surface Silver and Aluminum Mirrors. *J. Sol-Gel Sci. Technol.* **1997**, *8*, 451–457. [CrossRef]

20. Lugolole, R.; Obwoya, S.K. The Effect of Thickness of Aluminium Films on Optical Reflectance. *J. Ceram.* **2015**, *2015*, 1–6. [CrossRef]

21. Eder, A.; Schmid, G.H.S.; Mahr, H.; Eisenmenger-Sittner, C. Aspects of thin film deposition on granulates by physical vapor deposition. *Eur. Phys. J.* **2016**, *70*, 247. [CrossRef]

22. Schmid, G.H.S.; Eisenmenger-Sittner, C. A method for uniformly coating powdery substrates by magnetron sputtering. *Surf. Coat. Technol.* **2013**, *236*, 353–360. [CrossRef]

23. Park, J.-H. Chemical Vapor Deposition. In *Surface Engineering Series Volume 2*; ASM International: Almere, The Netherlands, 2001; pp. 1–3. ISBN 0-87170-731-4.

24. Lee, H.M.; Choi, S.-Y.; Kim, K.T.; Yun, J.-Y.; Jung, D.S.; Park, S.B.; Park, J. A Novel Solution-Stamping Process for Preparation of a Highly Conductive Aluminum Thin Film. *Adv. Mater.* **2011**, *23*, 5524–5528. [CrossRef] [PubMed]

25. Houghton, J. *The Physical of Atmopsheres*, 3rd ed.; Cambridge University Press: Cambridge, UK, 2002; xv +320p, ISBN 0-521-80456-6.

26. Brower, F.M.; Matzek, N.E.; Reigler, P.F.; Rinn, H.W.; Roberts, C.B.; Schmidt, D.L.; Snover, J.A.; Terada, K. Preparation and Properties of Aluminum Hydride. *J. Am. Chem. Soc.* **1976**, *98*, 2450. [CrossRef]

27. Vanyukhin, K.D.; Zakharchenko, R.V.; Kargin, N.I.; Pashkov, M.V. Study of structure and surface morphology of two-layer contact Ti/Al metallization. *Mod. Electron. Mater.* **2016**, *2*, 54–59. [CrossRef]

28. Das, N.; Islam, S. Design and Analysis of Nano-Structures Gratings for Conversion Efficiency Improvement in GaAs Solar Cells. *Energies* **2016**, *9*, 690. [CrossRef]

29. Moushumy, N.; Das, N.; Alameh, K.; Lee, Y.T. Design and development of silver nanoparticles to reduce the reflection loss of solar cells. In Proceedings of the 8th International Conference on High-capacity Optical Networks and Emerging Technologies, Riyadh, Saudi Arabia, 19–21 December 2011. [CrossRef]
30. Sharma, M.; Das, N.; Helwig, A.; Ahfock, T. Impact of Incident Light angle on the Conversion Efficiency of Nano-structured GaAs Solar Cells. In Proceedings of the Australasian Universities Power Engineering Conference (AUPEC), Melbourne, Australia, 19–22 November 2017. [CrossRef]

 *coatings*

Article

Study on β-Ga$_2$O$_3$ Films Grown with Various VI/III Ratios by MOCVD

Zeming Li [1], Teng Jiao [1], Daqiang Hu [1], Yuanjie Lv [2], Wancheng Li [1], Xin Dong [1,*], Yuantao Zhang [1], Zhihong Feng [2] and Baolin Zhang [1,*]

[1] State Key Laboratory on Integrated Optoelectronics, College of Electronic Science and Engineering, Jilin University, Qianjin Street 2699, Changchun 130012, China; zmli17@mails.jlu.edu.cn (Z.L.); jiaoteng18@mails.jlu.edu.cn (T.J.); hudq16@mails.jlu.edu.cn (D.H.); liwc@jlu.edu.cn (W.L.); zhangyt@jlu.edu.cn (Y.Z.)

[2] National Key Laboratory of Application Specific Integrated Circuit (ASIC), Hebei Semiconductor Research Institute, Shijiazhuang 050051, China; yuanjielv@163.com (Y.L.); ga917vv@163.com (Z.F.)

* Correspondence: dongx@jlu.edu.cn (X.D.); zbl@jlu.edu.cn (B.Z.)

Received: 20 March 2019; Accepted: 23 April 2019; Published: 26 April 2019

Abstract: β-Ga$_2$O$_3$ films were grown on sapphire (0001) substrates with various O/Ga (VI/III) ratios by metal organic chemical vapor deposition. The effects of VI/III ratio on growth rate, structural, morphological, and Raman properties of the films were systematically studied. By varying the VI/III ratio, the crystalline quality obviously changed. By decreasing the VI/III ratio from 66.9×10^3 to 11.2×10^3, the crystalline quality improved gradually, which was attributed to low nuclei density in the initial stage. However, crystalline quality degraded with further decrease of the VI/III ratio, which was attributed to excessive nucleation rate.

Keywords: β-Ga$_2$O$_3$; MOCVD; VI/III ratio

---

1. Introduction

β-Ga$_2$O$_3$, the most stable phase of Ga$_2$O$_3$, shows great potential because of its excellent material properties. It is a wide bandgap (WBG) semiconductor with band gap of ~4.9 eV, breakdown field of 8 MV cm^{-1} and Baliga's figure of merit of 3444 at room temperature, which offers more advantages in high-efficiency power device application than SiC and GaN [1]. Moreover, its high transparency in UV wavelength range, and excellent thermal and chemical stability also have great application potential in flat panel displays, UV detectors, and high-temperature gas sensors [2–6]. There are several ways to produce a β-Ga$_2$O$_3$ film, which include molecule beam epilayer (MBE) [7], metal organic chemical vapor deposition (MOCVD) [8], halide vapor phase epitaxy (HVPE) [4], chemical vapor deposition (CVD) [9], magnetron sputtering [10], and thermal oxidation [11]. Conventional CVD methods [12–14], especially MOCVD have several advantages, including excellent reproducibility and capability for scale-up to high-volume production [15]. Impressive studies on the growth of β-Ga$_2$O$_3$ by MOCVD have been recently reported. Lv et al. investigated the epitaxial relationship between β-Ga$_2$O$_3$ and sapphire substrates [16]. Zhuo et al. studied the control of the crystal phase composition of the Ga$_2$O$_3$ thin film [17]. Sbrockeyet al. demonstrated the large-area growth of β-Ga$_2$O$_3$ films using rotating disc MOCVD reactor technology [15]. Alema et al. studied the growth rates of β-Ga$_2$O$_3$ epitaxial films by close coupled showerhead MOCVD [18]. Takiguchi et al. studied β-Ga$_2$O$_3$ epitaxial films obtained by low temperature MOCVD [19]. Chen et al. investigated the effect of growth pressure on the characteristics of β-Ga$_2$O$_3$ films grown on GaAs (100) substrates [20]. However, the crystalline quality of heteroepitaxial β-Ga$_2$O$_3$ films has not been able to meet the requirements of device fabrication so far.

In this paper, β-Ga$_2$O$_3$ films were grown by MOCVD on sapphire (0001) substrates with various VI/III ratios. In addition, the effects of VI/III ratio on growth rate, structural, morphological, and Raman properties were systematically studied. By varying the ratio, the crystalline quality of the films was effectively improved.

2. Materials and Methods

2.1. Materials

High purity O_2 (purity, 5 N) and trimethylgallium (TMGa, 6 N in purity, Nata Opto-electronic Material Co., Nanjing, China) and were used as oxidant and organometallic source, respectively. High purity Ar (purity, 6 N) worked as a carrier gas.

2.2. Preparation

The β-Ga_2O_3 films were grown on sapphire (0001) substrates by MOCVD. The equipment was modified from an Emcore D180 MOCVD (Emcore, Alhambra, CA, USA). The close coupled showerhead method is used; the highest growth temperature of the MOCVD was 1150 °C. Before the growth process, the substrates were cleaned sequentially by acetone, ethanol, deionized water in an ultrasonic bath, and then dried with N_2. The growth pressure and substrate temperature were kept at 20 mbar and 750 °C during the whole growth process, respectively. High purity O_2 was injected into the reaction chamber with a fixed flow rate of 1200 sccm. TMGa was stored in a stainless steel bubbler, maintained at 1 °C. The pressure inside the bubbler were kept at 900 Torr. Ar carrier gas passed through the TMGa bubbler and delivered the TMGa vapor to the reactor. To obtain β-Ga_2O_3 films grown with various VI/III ratios, the flow rates of Ar carrier gas were varied from 5 sccm to 60 sccm (5 sccm, 15 sccm, 30 sccm, 45 sccm, 60 sccm). The growth time was 30 min.

2.3. Characterization

The structural properties of β-Ga_2O_3 films were investigated by X-ray diffractometer (XRD, Rigaku, Ultima IV, Tokyo, Japan, $\lambda = 0.15406$ nm, graphite filter). The morphological properties of the β-Ga_2O_3 films were studied by field emission scanning electron microscopy (FESEM, JSM-7610F, JEOL, Tokyo, Japan) and atomic force microscopy (AFM, Veeco, Plainview, NY, USA). Raman properties of the films was analyzed by a Raman spectrometer (HORIBA, LABRAM HR EVO, Kyoto, Japan) using a wavelength of $\lambda = 633$ nm laser. The thicknesses of the films were measured by a thin film analyzer (F40, Filmetrics, San Diego, CA, USA).

3. Results and Discussion

The molar flow rates in the experiments can be calculated by Equations (1)–(3) [21,22]:

$$\ln(P_{MO}) = a - b/T \tag{1}$$

where P_{MO} is the vapor pressure of TMGa, $a = 8.07$, $b = 1703$, T is the thermodynamic temperature of TMGa,

$$n_{MO} = F \times P_{MO}/[V_m \times (P_{bub} - P_{MO})], \tag{2}$$

where n_{MO} is the molar flow rate of TMGa, F is the flow rate of carrier gas, $V_m = 22414$ cm^3/mol, P_{bub} is the pressure inside the bubbler,

$$n_O = F_O/V_m, \tag{3}$$

where n_O is the molar flow rate of O_2, F_O is the flow rate of O_2. The VI/III ratios in the experiments are shown in Table 1.

Table 1. The VI/III ratios at various flow rates of Ar carrier gas.

Flow Rate for Ar Carrier Gas (sccm)	VI/III ratio ($\times 10^3$)
5	66.9
15	22.3
30	11.2
45	7.4
60	5.6

3.1. Growth Rate Analysis

To investigate the growth rates, the thicknesses of the samples were measured by a thin film analyzer. The sample obtained with VI/III ratio of 5.6×10^3 is unsuitable for such analysis due to its excessively rough surface [18]. The growth rate showed a strong dependence on the VI/III ratio (Figure 1). Because the flow rate of oxygen was a constant, the growth rate was mainly limited by the flow rate of organometallic source. By increasing the flow rates of Ar carrier gas from 5 sccm to 45 sccm, the VI/III ratio decreased from 66.9×10^3 to 7.4×10^3, and the growth rate improved from 0.26 to 1.98 μm/h.

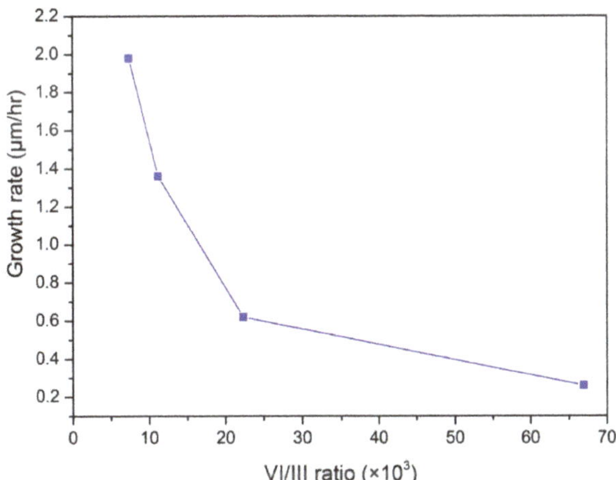

Figure 1. Growth rates of the samples obtained with various VI/III ratios.

3.2. XRD Analysis

Figure 2 shows the XRD θ–2θ scan patterns of β-Ga_2O_3 films grown with various VI/III ratios. For the film grown with VI/III ratio of 66.9×10^3, except the diffraction peaks of Al_2O_3 substrate, only three peaks located at 18.76°, 38.10° and 58.84° could be observed, which related to β-Ga_2O_3 (-201), (-402), and (-603). It indicated that the thin film consisted of pure β-Ga_2O_3. By decreasing the VI/III ratio from 66.9×10^3 to 11.2×10^3, the three peaks of β-Ga_2O_3 were strengthened and sharpened. The crystallite sizes along the direction vertical to (-201) plane of the samples obtained with the VI/III ratios of 66.9×10^3, 22.3×10^3, and 11.2×10^3 were calculated to be 11.2, 12.2, and 17.5 nm, respectively (by Scherrer equation). Larger crystallite sizes indicated lower defect density and an improvement of crystalline quality. Lower VI/III ratio was helpful to reduce the nuclei density in the initial stage of deposition process and enlarge the size of islands in the subsequent stage, which indicated that less defects occurred in island coalescence [23,24]. However, further decreasing the VI/III ratio caused crystalline quality degradation. For the film grown with VI/III ratio of 7.4×10^3, the intensities of the three β-Ga_2O_3 peaks declined, and peaks related to β-Ga_2O_3 (401), (-601), (601), and (-801) were observed, indicating the polycrystalline structure of the film. The change in crystalline structure is caused by excessive nucleation rate with this VI/III ratio. At this nucleation rate, the deposited particles were unable to migrate to the appropriate lattice positions, and the films grew and oriented in unsuitable directions, which caused random growth. As for the sample obtained with VI/III ratio of 5.6×10^3, the change in crystalline structure was obvious—15 peaks of β-Ga$_2$O$_3$ showed up. The crystallite sizes of the films grown with VI/III ratio of 7.4×10^3 and 5.6×10^3 were calculated to be 14.2 and 21.3 nm, respectively.

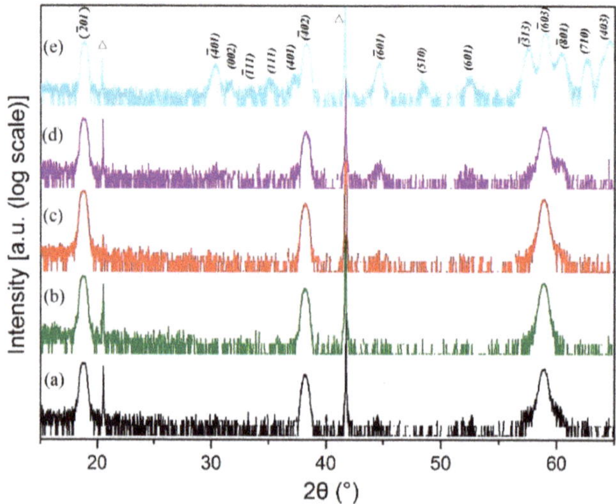

Figure 2. XRD θ–2θ scan patterns of β-Ga$_2$O$_3$ films grown with various VI/III ratios: (**a**) 66.9×10^3; (**b**) 22.3×10^3; (**c**) 11.2×10^3; (**d**) 7.4×10^3; (**e**) 5.6×10^3. Δ the peaks of the sapphire substrates.

3.3. AFM Analysis

To investigate the effects of VI/III ratios on the surface morphology of β-Ga$_2$O$_3$ films, AFM was carried out; the images are shown in Figure 3. The surface roughness of the films depended highly on the VI/III ratios. For the film grown with the VI/III ratios from 66.9×10^3 to 11.2×10^3, root-mean-square (RMS) surface roughness increased from 3.71 to 7.83 nm. The hillocks on the surfaces enlarged and decreased in density, in good agreement with the XRD analysis. By decreasing the VI/III ratio to 7.4×10^3, the surface roughness had little change, while the morphology changed greatly. Many wheat-like structures were observed, which means that excessive nucleation rate hindered particle migration and caused random growth. For the film grown with VI/III ratio of 5.6×10^3, the roughness increased greatly, even reaching 56.3 nm (seven times higher than that of any other film), in accordance with its XRD pattern (Figure 2).

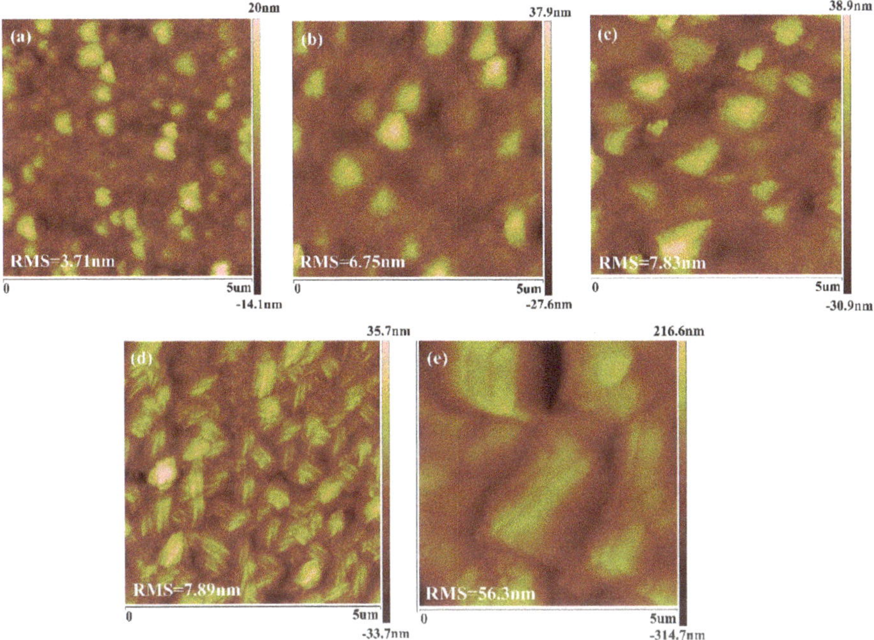

Figure 3. AFM images (5 μm × 5 μm) of β-Ga$_2$O$_3$ films grown with various VI/III ratios: (**a**) 66.9 × 10^3; (**b**) 22.3 × 10^3; (**c**) 11.2 × 10^3; (**d**) 7.4 × 10^3; (**e**) 5.6 × 10^3.

3.4. FESEM Analysis

According to XRD analysis of all the films, the sample obtained with VI/III ratio of 11.2 × 10^3 was measured by FESEM. Figure 4 shows the top and cross-sectional views of FESEM images of the sample. The surface with minor defects is in accordance with the AFM image in Figure 3. The relatively smooth cross-sectional image indicates high film quality. In addition, the thickness measured by the cross-sectional view images is about 0.68 μm.

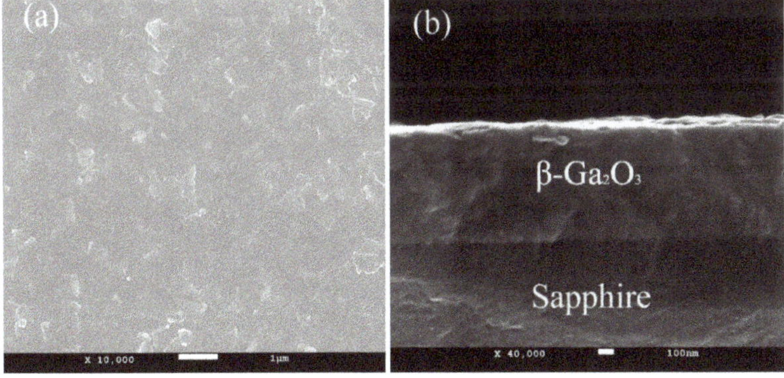

Figure 4. FESEM images of β-Ga$_2$O$_3$ films grown with VI/III ratio of 11.2 × 10^3: (**a**) Top-view; (**b**) cross-sectional view.

3.5. Raman Analysis

Figure 5 presents the Raman spectra of β-Ga$_2$O$_3$ films grown with various VI/III ratios. For comparison, the Raman spectra of the sapphire substrates is also shown in this figure. Except for the peaks related to the substrates, only one Raman peak related to β-Ga$_2$O$_3$ was observed. For the film grown with VI/III ratio of 66.9 × 10^3, due to poor crystalline quality and a smooth surface, only one peak related to β-Ga$_2$O$_3$ was clearly observed. By decreasing the VI/III ratio, due to the change in crystalline quality and roughness, more peaks related to β-Ga$_2$O$_3$ showed up, which were gradually enhanced. However, when the VI/III ratio was decreased to 5.6 × 10^3, owing to the excessively rough surface of the obtained sample, its surface area increased and its Raman spectrum changed greatly. Ten peaks related to β-Ga$_2$O$_3$ showed up and the peaks were enhanced greatly. The 10 peaks were divided into three categories [25–27]—the peaks located at 115, 147, 171, and 201 cm^{-1} were attributed to libration and translation of tetrahedral-octahedra chains; the peaks located at 322, 349, and 476 cm^{-1} were attributed to deformation of GaO$_6$ octahedra, and the peaks located at 631, 655, and 768 cm^{-1} were attributed to stretching and bending of GaO$_4$ tetrahedra. The Raman results confirmed that all the obtained films consisted of pure β-Ga$_2$O$_3$.

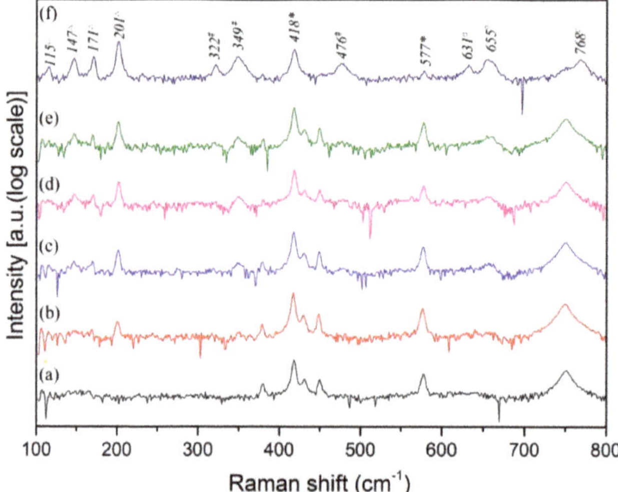

Figure 5. Raman spectra of β-Ga$_2$O$_3$ films grown with various VI/III ratios: (**a**) sapphire substrate; (**b**) 66.9 × 10^3; (**c**) 22.3 × 10^3; (**d**) 11.2 × 10^3; (**e**) 7.4 × 10^3; (**f**) 5.6 × 10^3. *, the Raman peaks related to sapphire. Δ, the Raman peaks related to tetrahedral-octahedra chains. #, the Raman peaks related to GaO$_6$ octahedra. □, the Raman peaks related to GaO$_4$ tetrahedra.

4. Conclusions

In summary, β-Ga$_2$O$_3$ films were grown on sapphire (0001) substrates with various VI/III ratios by MOCVD. By varying the VI/III ratio, the crystalline quality obviously changed. For the film grown with VI/III ratios from 66.9 × 10^3 to 11.2 × 10^3, the crystalline quality improved gradually, attributed to low nuclei density in the initial stage. However, further decreasing the VI/III ratio caused degradation of crystalline quality, and the morphological and Raman properties changed greatly, which was attributed to excessive nucleation rate. This work offers a feasible way to improve the crystalline quality of heteroepitaxial β-Ga$_2$O$_3$ films and is beneficial for device fabrication.

Author Contributions: Conceptualization, Z.L.; Methodology, W.L.; Validation, D.H.; Formal Analysis, D.H.; Investigation, T.J.; resources, B.Z.; Data Curation, Y.Z.; Writing—Original Draft Preparation, Z.L.; Writing—Review and Editing, X.D.; Visualization, Y.L.; Supervision, Z.F.; Project Administration, X.D.; Funding Acquisition, X.D.

Funding: This research was funded by the National Natural Science Foundation of China, Grant Numbers 61774072, 61376046, 61674068 and 61404070; the Science and Technology Developing Project of Jilin Province, Grant Number 20170204045GX, 20150519004JH, 20160101309JC; the National Key Research and Development Program, Grant Number 2016YFB0401801; and the Program for New Century Excellent Talents in University, Grant Number NCET-13-0254.

Conflicts of Interest: The authors declare no conflict of interest.

References

1. Higashiwaki, M.; Sasaki, K.; Murakami, H.; Kumagai, Y.; Koukitu, A.; Kuramata, A.; Masui, T.; Yamakoshi, S. Recent progress in Ga_2O_3 power devices. *Semicond. Sci. Technol.* **2016**, *31*, 034001. [CrossRef]

2. Goyal, A.; Yadav, B.S.; Thakur, O.P.; Kapoor, A.K.; Muralidharan, R. Effect of annealing on β-Ga_2O_3 film grown by pulsed laser deposition technique. *J. Alloy. Compd.* **2014**, *583*, 214–219. [CrossRef]

3. Kang, H.C. Heteroepitaxial growth of multidomain Ga_2O_3/sapphire (001) thin films deposited using radio frequency magnetron sputtering. *Mater. Lett.* **2014**, *119*, 123–126. [CrossRef]

4. Nikolaev, V.I.; Pechnikov, A.I.; Stepanov, S.I.; Nikitina, I.P.; Smirnov, A.N.; Chikiryaka, A.V.; Sharofidinov, S.S.; Bougrov, V.E.; Romanov, A.E. Epitaxial growth of (-201) β-Ga_2O_3 on (0001) sapphire substrates by halide vapour phase epitaxy. *Mater. Sci. Semicond. Process* **2016**, *47*, 16–19. [CrossRef]

5. Liu, X.Z.; Guo, P.; Sheng, T.; Qian, L.X.; Zhang, W.L.; Li, Y.R. β-Ga_2O_3 thin films on sapphire pre-seeded by homo-self-templated buffer layer for solar-blind UV photodetector. *Opt. Mater.* **2016**, *51*, 203–207. [CrossRef]

6. Kumar, S.; Tessarek, C.; Christiansen, S.; Singh, R. A comparative study of β-Ga_2O_3 nanowires grown on different substrates using CVD technique. *J. Alloy. Compd.* **2014**, *587*, 812–818. [CrossRef]

7. Sasaki, K.; Higashiwaki, M.; Kuramata, A.; Masui, T.; Yamakoshi, S. MBE grown β-Ga_2O_3 and its power device applications. *J. Cryst. Growth* **2013**, *378*, 591–595. [CrossRef]

8. Du, X.; Mi, W.; Luan, C.; Li, Z.; Xia, C.; Ma, J. Characterization of homoepitaxial β-Ga_2O_3 films prepared by metal–organic chemical vapor deposition. *J. Cryst. Growth* **2014**, *404*, 75–79. [CrossRef]

9. Terasako, T.; Kawasaki, Y.; Yagi, M. Growth and morphology control of β-Ga_2O_3 nanostructures by atmospheric-pressure CVD. *Thin Solid Films* **2016**, *620*, 23–29. [CrossRef]

10. Ogita, M.; Saika, N.; Nakanishi, Y.; Hatanaka, Y. β-Ga_2O_3 thin films for high-temperature gas sensors. *Appl. Surf. Sci.* **1999**, *142*, 188–191. [CrossRef]

11. Chen, P.; Zhang, R.; Xu, X.F.; Zhou, Y.G.; Chen, Z.Z.; Xie, S.Y.; Li, W.P.; Zheng, Y.D. Oxidation of gallium nitride epilayers in dry oxygen. *Mater. Res. Soc. Internet J. Nitride Semicond. Res.* **2000**, *5*, 866–872. [CrossRef]

12. Komiyama, H.; Shimogaki, Y.; Egashira, Y. Chemical reaction engineering in the design of CVD reactors. *Chem. Eng. Sci.* **1999**, *54*, 1941–1957. [CrossRef]

13. Parikh, R.P.; Adomaitis, R.A. An overview of gallium nitride growth chemistry and its effect on reactor design: Application to a planetary radial-flow CVD system. *J. Cryst. Growth* **2006**, *286*, 259–278. [CrossRef]

14. de Graaf, A.; van Deelen, J.; Poodt, P.; van Mol, T.; Spee, K.; Grob, F.; Kuypers, A. Development of atmospheric pressure CVD processes for high quality transparent conductive oxides. *Energy Procedia* **2010**, *2*, 41–48. [CrossRef]

15. Sbrockey, N.M.; Salagaj, T.; Coleman, E.; Tompa, G.S.; Moon, Y.; Kim, M.S. Large-area MOCVD growth of Ga_2O_3 in a rotating disc reactor. *J. Electron. Mater.* **2015**, *44*, 1357–1360. [CrossRef]

16. Lv, Y.; Ma, J.; Mi, W.; Luan, C.; Zhu, Z.; Xiao, H. Characterization of β-Ga_2O_3 thin films on sapphire (0001) using metal-organic chemical vapor deposition technique. *Vacuum* **2012**, *86*, 1850–1854. [CrossRef]

17. Zhuo, Y.; Chen, Z.; Tu, W.; Ma, X.; Pei, Y.; Wang, G. β-Ga_2O_3 versus ε-Ga_2O_3: Control of the crystal phase composition of gallium oxide thin film prepared by metal-organic chemical vapor deposition. *Appl. Surf. Sci.* **2017**, *420*, 802–807. [CrossRef]

18. Alema, F.; Hertog, B.; Osinsky, A.; Mukhopadhyay, P.; Toporkov, M.; Schoenfeld, W.V. Fast growth rate of epitaxial β-Ga_2O_3 by close coupled showerhead MOCVD. *J. Cryst. Growth* **2017**, *475*, 77–82. [CrossRef]

19. Takiguchi, Y.; Miyajima, S. Effect of post-deposition annealing on low temperature metalorganic chemical vapor deposited gallium oxide related materials. *J. Cryst. Growth* **2017**, *468*, 129–134. [CrossRef]

20. Chen, Y.; Liang, H.; Xia, X.; Liu, Y.; Luo, Y.; Du, G. Effect of growth pressure on the characteristics of β-Ga_2O_3 films grown on GaAs (100) substrates by MOCVD method. *Appl. Surf. Sci.* **2015**, *325*, 258–261. [CrossRef]

21. Stringfellow, G.B. *Organometallic Vapor-Phase Epitaxy: Theory and Practice*, 2nd ed.; Academic Press: San Diego, CA, USA, 1999; pp. 163–169.

22. Lu, D.; Duan, S. *Metalorganic Vapor Phase Epitaxy Foundation and Application*, 1st ed.; Science Press: Beijing, China, 2009; pp. 11–46.
23. Vennegues, P.; Beaumont, B.; Haffouz, S.; Vaille, M.; Gibart, P. Influence of in situ sapphire surface preparation and carrier gas on the growth mode of GaN in MOVPE. *J. Cryst. Growth* **1998**, *187*, 167–177. [CrossRef]
24. Yang, T.; Uchida, K.; Mishima, T.; Kasai, J.; Gotoh, J. Control of initial nucleation by reducing the V/III ratio during the early stages of GaN growth. *Phys. Stat. Sol. A* **2000**, *180*, 45–50. [CrossRef]
25. Kumar, S.; Sarau, G.; Tessarek, C.; Bashouti, M.Y.; Hähnel, A.; Christiansen, S.; Singh, R. Study of iron-catalysed growth of β-Ga$_2$O$_3$ nanowires and their detailed characterization using TEM, Raman and cathodoluminescence techniques. *J. Phys. D Appl. Phys.* **2014**, *47*, 435101. [CrossRef]
26. Rao, R.; Rao, A.M.; Xu, B.; Dong, J.; Sharma, S.; Sunkara, M.K. Blue shifted Raman scattering and its correlation with the [110] growth direction in gallium oxide nanowires. *J. Appl. Phys.* **2005**, *98*, 094312. [CrossRef]
27. Gao, Y.H.; Bando, Y.; Sato, T.; Zhang, Y.F.; Gao, X.Q. Synthesis, Raman scattering and defects of β-Ga$_2$O$_3$ nanorods. *Appl. Phys. Lett.* **2002**, *81*, 2267–2269. [CrossRef]

 *coatings*

Article

Structure and Conductivity Studies of Scandia and Alumina Doped Zirconia Thin Films

Mantas Sriubas [1], Nursultan Kainbayev [1,2], Darius Virbukas [1], Kristina Bočkutė [1,*], Živilė Rutkūnienė [1] and Giedrius Laukaitis [1]

[1]   Physics Department, Kaunas University of Technology, Studentu str. 50, LT-51368 Kaunas, Lithuania; mantas.sriubas@ktu.lt (M.S.); nursultan.kainbayev@ktu.edu (N.K.); darius.virbukas@ktu.lt (D.V.); zivile.rutkuniene@ktu.lt (Ž.R.); giedrius.laukaitis@ktu.lt (G.L.)
[2]   Department of Thermal Physics and Technical Physics, Al-Farabi Kazakh National University, 71 Al-Farabi Ave., 050040 Almaty, Kazakhstan
*   Correspondence: kristina.bockute@ktu.lt

Received: 22 March 2019; Accepted: 8 May 2019; Published: 12 May 2019

Abstract: In this work, scandia-doped zirconia (ScSZ) and scandia–alumina co-doped zirconia (ScSZAl) thin films were prepared by electron beam vapor deposition. X-ray diffraction (XRD) results indicated a presence of ZrO_2 cubic phase structure, yet Raman analysis revealed the existence of secondary tetragonal and rhombohedral phases. Thus, XRD measurements were supported by Raman spectroscopy in order to comprehensively analyze the structure of formed ScSZ and ScSZAl thin films. It was also found that Al dopant slows down the formation of the cubic phase. The impedance measurements affirmed the correlation of the amount of secondary phases with the conductivity results and nonlinear crystallite size dependence.

Keywords: scandium stabilized zirconia thin films; e-beam physical vapor deposition; thin films ceramics; Raman spectroscopy; X-ray diffraction

---

1. Introduction

Zirconium oxide-based materials have been extensively studied for a few decades as electrolytes and fulfill almost all requirements: high oxide ion conductivity (0.1–0.01 S/cm), low electronic conductivity, chemical stability, mechanical strength, and low cost [1].

Pure ZrO_2, due to its low number of vacancies, has a low specific ionic conductivity and has a monoclinic structure, and up to 1170 °C does not exhibit high ionic conductivity. The highly-conductive cubic phase forms only at 2370 °C [2]. A tetragonal phase exists in the middle-temperature range (1170–2370 °C). The monoclinic–tetragonal (m → t) phase transition is produced by a condensation of two phonons at point M of the Brillouin zone of the tetragonal phase (P42/nmc) and the tetragonal–cubic (t → c) phase transition is produced by a condensation of phonons at point X of the Brillouin zone of the cubic phase (Fm3m) [3,4].

To improve ionic conductivity and stabilize the cubic lattice at low temperatures, zirconia is doped with trivalent or bivalent elements, such as Ca, Mg, Y, and Sc, etc. [5,6]. The stabilization process can be explained by crystal chemistry, i.e., oxygen vacancies associated with Zr can provide stability for cubic zirconia. Vacancies introduced by the oversized dopants are located as the nearest neighbors to the Zr atoms, leaving the eightfold coordination to dopant cations. Undersized dopants compete with Zr ions for oxygen vacancies in zirconia, resulting in six-fold oxygen coordination and a large disturbance to the surrounding next nearest neighbors [7–9].

Scandia-doped zirconia (ScSZ) is considered the most suitable material at intermediate temperatures because it has much higher ionic conductivity than yttria-doped zirconia [10]. However, four different phases can be observed below 600 °C for ScSZ systems depending on the concentration of Sc_2O_3, i.e., cubic-tetragonal (5–7 mol %), cubic (8–9 mol %), and rhombohedral (10–15 mol %) [7,11]. Transitions between the highly conductive cubic phase and tetragonal or rhombohedral phases cause an abrupt decrease in ionic conductivity and suspend the application of ScSZ for intermediate temperature solid oxide fuel cells (IT-SOFC) [7,12]. Co-doping can be a possible solution. Adding dopants like CeO_2, Bi_2O_3, Al_2O_3, and etc., has already been attempted [13–15]. Results vary for different dopants. Mostly, the cubic phase has been mixed with other phases (rhombohedral, monoclinic, or tetragonal) except for co-doping 0.5 mol % of Al_2O_3. It has been demonstrated that 0.5 mol % of Al_2O_3 stabilizes the cubic phase of ScSZ to room temperature [13]. The mechanism of cubic phase stabilization is not well known. It has been suggested that the strain in the crystalline, which is induced by adding a secondary dopant with a different radius to the Sc^{3+}, is involved [13,16]. However, this research was based only on the X-ray diffraction (XRD) method, which cannot fully prove the existence of a fully stabilized cubic structure. In a mixture of phases, XRD peaks of the main cubic phase and secondary phases (tetragonal and rhombohedral phases) can overlap and the presence of secondary phases is not reflected ine XRD analysis [17], indicating that XRD analysis should be supported by an additional characterization method, e.g., Raman spectroscopy. Raman spectroscopy is a powerful analysis tool used to study the fundamental vibrational characteristics of molecules and is very sensitive to phase changes in the material. It can provide information about the phase, the chemical composition of the material, and oxygen vacancies [18,19]. Light scattering occurs and a small percentage of the scattered light may be shifted in frequency when monochromatic light is incident on the material. This frequency shift of the Raman scattered light is directly related to the structural properties of the material. The change in the phonon frequency of the vibrational mode will be produced in the presence of discontinuation of translational symmetry in the crystalline material due to the doping and secondary phases. It is known that the wavenumber of the vibrational modes follows a linear relationship with chemical composition as well as with strain induced in the crystalline lattice [20]. In the case of phase transformation of ZrO_2, Raman spectroscopy can determine the change of the bond length and the angle between cation and anion [21]. Based on the reports of other authors, monoclinic [12,18,22], tetragonal [15,21–24], rhombohedral [12,24], and cubic [12,15,18,24] phases can be identified from Raman spectra for ZrO_2. Moreover, the great majority of investigations analyzing co-doping of ScSZ have been focused on the powders and pellets. The properties and phase transitions of co-doped ScSZ thin films have not been studied enough. Thin films can be formed using a variety of deposition techniques, such as screen printing [25], wet powder spraying [26], and electrostatic spray deposition [27], etc. Electron beam vapor deposition allows producing a dense and homogenous thin film with a strictly controllable thin film growing process during the deposition [28]. Furthermore, it is a particularly appropriate formation method for ceramic thin films, which are distinguished by their high melting temperature. In e-beam evaporation, Sc-doped and Sc and Al co-doped ZrO_2 evaporate by partial dissociation [29–31], causing the different structure and crystallographic phases of the formed thin films than powders. In this paper, the structure and ionic conductivity of thin ScSZ and scandium alumina stabilized zirconia films (ScAlSZ) are analyzed using XRD, energy dispersive X-ray spectrsoscopy (EDS), Raman, and electrochemical impedance spectroscopy (EIS) methods.

2. Materials and Methods

ScSZ and ScAlSZ thin films were formed using an e-beam physical vapor deposition system (Kurt J. Lesker EB-PVD 75, Hastings, UK). The formation was carried out on crystalline Alloy 600 (Fe-Ni-Cr) and polycrystalline Al_2O_3 substrates using deposition rates from 0.2 to 1.6 nm/s in steps of 0.2 nm/s. The thickness (~1500 nm) and deposition rate were controlled with an INFICON (Inficon, Bad Ragaz, Switzerland) crystal sensor. The temperature of substrates was changed from room temperature (20 °C) to 600 °C temperature. In order to get a homogenous thin film, substrates were rotated at 8 rpm speed. The acceleration voltage of the electron gun was kept at a constant of 7.9 kV and the required deposition rate was achieved by adjusting the e-beam current in the range of 60–100 mA. An initial pressure of ~2.0×10^{-4} Pa and working pressure of ~2.0×10^{-2} Pa was used in the vacuum chamber during the experiments. The substrates were cleaned in an ultrasonic bath and treated in Ar$^+$ ion plasma for 10 min before deposition. The powders of $(Sc_2O_3)_{0.10}(ZrO_2)_{0.90}$ (ScSZ) and $(Sc_2O_3)_{0.10}(Al_2O_3)_{0.01}(ZrO_2)_{0.89}$ (ScSZAl) (Nexceris, LLC, Fuelcellmaterials, Lewis Center, OH, USA) were pressed into pellets and used as evaporating material. Pellets were evaporated using a single e-beam configuration. Elemental analysis was performed using energy-dispersive X-ray spectroscopy (EDS, BrukerXFlash QUAD 5040, Bruker AXS GmbH, Karlsruhe, Germany) and the atomic ratios of Sc, Al, and Zr are presented in Table 1.

Table 1. Atomic ratios of Sc, Al, and Zr in the initial powders.

Initial Powders	c_{Sc}, %	c_{Al}, %	c_{Zr}, %
ScSZ	0.19	–	0.81
ScAlSZ	0.19	0.02	0.79

The crystallographic nature of the formed thin films was determined using XRD (D8 Discover (Bruker AXS GmbH, Billerica, MA, USA) standard Bragg focusing geometry with Cu Kα_1 (λ = 0.154059 nm) radiation, 0.01° step, and Lynx eye PSD detector). Rietveld analysis was performed to calculate ratios of cubic and rhombohedral phases in the ScSZAl thin films. The refinement was conducted with isotropic atomic thermal parameters [32]. The size of the crystallites and lattice constants were estimated with the software Topas 4-1. The Pawley method was used to fit XRD patterns and the crystallite size was estimated using the Scherrer equation [33].

Raman scattering measurements were performed using the Raman microscope inVia (Renishaw, Gloucestershire, UK). The excitation beam from a diode laser of 532 nm wavelength was focused on the sample using a 50× objective (NA = 0.75, Leica, Wetzlar, Germany) and the diameter of the laser spot was 4 µm. Laser power at the sample surface varied from 1.75 to 3.5 mW. The integration time was 15–30 s and the signal was accumulated five times and then averaged. The Raman Stokes signal was dispersed with a diffraction grating (2400 grooves/mm) and data was recorded using a Peltier cooled charge-coupled device (CCD) detector (Renishaw, Gloucestershire, UK) (1024 × 256 pixels). This system yields a spectral resolution of about 1 cm^{-1}. Silicon was used to calibrate the Raman setup for both the Raman wavenumber and spectral intensity. Positions of Raman peaks were determined by fitting the data to the Lorentz line shape using a peak fit option in OriginPro 2016 software. The phase ratio was calculated using the formula [18]

$$\vartheta_C = \frac{I_C}{I_C + I_t + I_\beta} \tag{1}$$

where I_c, I_t, and I_β are the scattering intensity of cubic (~620 cm^{-1}), tetragonal (~473 cm^{-1}), and rhombohedral (~557 cm^{-1}) Raman modes, respectively.

Electrical characterization and impedance spectroscopy measurements were performed using a Probostat$^®$ (NorECs AS, Oslo, Norway) measurement cell in the frequency range 1–10^6 Hz and a 200–1000 °C temperature interval. Electrodes of the geometry 3 mm × 10 mm ($L \times B$) were made of Pt ink and applied on top of the thin films using a mask reproducing the geometry of the electrodes (the two-probe method). The total conductivity was calculated according to:

$$\sigma = \frac{L_e}{R_s A} = \frac{L_e}{R_s h l_e} \tag{2}$$

where L_e is the distance between the Pt electrodes, R_s is the resistance obtained from the impedance spectra, A is the cross-sectional area, h is the thickness of the thin films, and l_e is the length of the electrodes.

3. Results

XRD measurements revealed that ScSZ and ScSZAl powders have a mixture of ZrO_2 rhombohedral (JCPSD No. 01-089-5482) and monoclinic (JCPSD No. 01-089-5474) phases (Figure 1a). It can be seen that the prevailing phase is rhombohedral, showing sharp (003), (101), (012), (104), (110), (015), (11-3), (021), (006), and (202) peaks. Raman spectra of investigated powders are in good agreement with the XRD results (Figure 1b). The peaks detected at 147, 148, 176, 190, 191, 259, 260, 315, 321, 384, 386, 422, 471, 473, 507, 517, 552, 554, 585, 597, and 640 cm^{-1}, and around 800–1100 cm^{-1} showing a polymorphism of ZrO_2, can be assigned to monoclinic, tetragonal, and rhombohedral phases. The peaks at 507–517, 552–554, 585, and 597 cm^{-1} belong to the β-rhombohedral phase [12], the peaks at 147–148, 259–260, 315–321, 384–386, 422, 547, 551, and 640 cm^{-1} are indicative of the tetragonal phase and the peaks at 176, 190, 191, and 469–473 cm^{-1} belong to the monoclinic phase [18,22]. Peaks observed above 800 cm^{-1} correspond to the second-order active Raman modes wave numbers combination [34]. The Raman spectra of cubic zirconia should consist of a weak broad line peak assigned to a single Raman mode F2g symmetry centered between 605 and 630 cm^{-1}. Peaks corresponding to the cubic phase of ZrO_2 were not observed in Raman spectra of the powders [12,15,18,21,24]. The obtained results agree with the results of other authors [24,35]. The material should have a rhombohedral structure if the Sc_2O_3 concentration is above 9% [5,11]. Raman peaks corresponding to the tetragonal phase (Figure 1b) indicate that the tetragonal structure is deformed and is predominant mostly at the grain boundary [15].

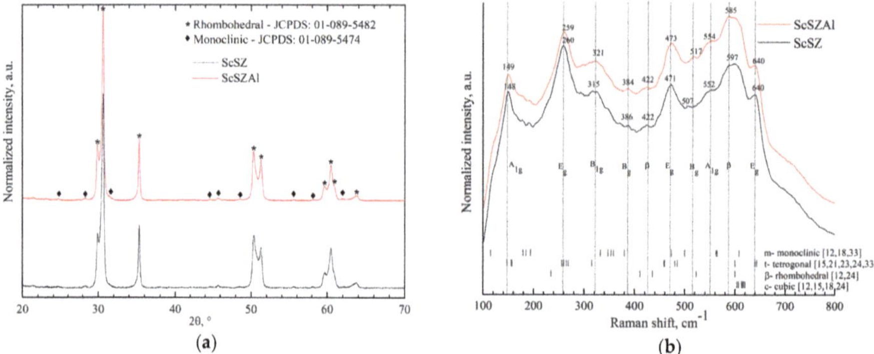

Figure 1. ScSZ and ScSZAl powders: (**a**) X-ray diffraction (XRD) patterns and (**b**) Raman shifts.

It is known that ZrO_2 evaporates by partial decomposition during e-beam evaporation process [29,31]:

$$ZrO_{2(s)} = ZrO_{2(g)} \tag{3}$$

$$ZrO_{2(s)} = ZrO_{(g)} + O_{(g)} \tag{4}$$

$$ZrO_{2(s)} = Zr_{(g)} + O_{2(g)} \tag{5}$$

Therefore, the vapor phase of ScSZ and ScSZAl powders could consist of ZrO_2, ZrO, O_2, O, Sc_2O_3, Sc_2O_2, ScO, Sc, Al_2O_3, Al_2O_2, AlO, Al atoms, molecules, and molecule fragments. The atoms and molecule fragments landed on the surface of the substrate migrate and form grains. The grains with the densest planes are selected, e.g., (111) for face-centered cubic. However, surface diffusivity plays an important role in the formation of textured thin films. Adatoms deposited near grain boundaries have a higher probability of becoming incorporated at a low diffusivity surface in comparison to adatoms at high diffusivity planes having longer mean free paths with correspondingly higher probabilities of moving off the plane and becoming trapped on the adjacent grains [36]. Therefore, grains with low surface diffusivities grow faster. A similar growth mechanism was observed in the ScSZ and ScSZAl thin films. The thin films have a cubic structure (Figure 2a) and the positions of (111), (200), (220),

(311), and (222) orientation peaks in the XRD patterns (measured at room temperature) confirm this (JCPSD No. 01-089-5483). The preferential orientation is (200) at low temperature (up to 450 °C) and preferential orientation is (111) at high temperatures because adatoms have high enough diffusion energy to move on the surface and become trapped on the high diffusivity surfaces. Moreover, the positions of the peaks of the ScSZAl thin films are shifted to higher angles by 0.2° in comparison to the positions of the peaks of the ScSZ thin films (Figure 2b) due to distortions in the crystal lattice. The ion of Al (~0.53 Å) is smaller than the ion of zirconium (0.84 Å).

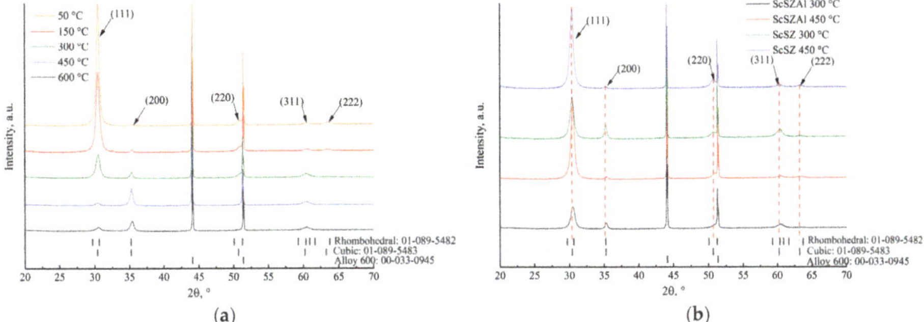

Figure 2. XRD patterns (measured at room temperature) of (**a**) ScSZAl thin films and (**b**) ScSZ and ScSZAl thin films deposited on Alloy 600 substrates using a 0.4 nm/s deposition rate.

However, Rietveld analysis showed that a mixture of cubic (87.8%) and rhombohedral (12.2%) phases can be found in the ScSZAl thin films (Figure 3) with equal probability as a pure cubic phase. The weighted profile R-factors (Rwp) are almost the same for pure cubic (1.45) and the mixture of cubic and rhombohedral (1.35). Therefore, the peaks of cubic and rhombohedral phases can be overlapped [17,35]. In this case, XRD patterns require a complimentary analysis and Raman spectroscopy can be a quick solution [15,21].

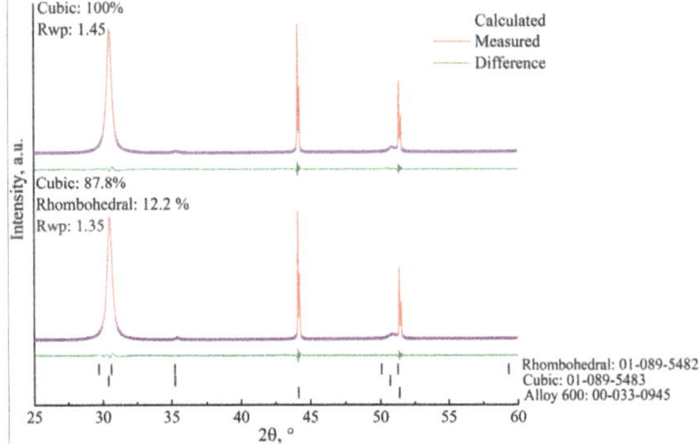

Figure 3. Rietveld analysis of XRD patterns of ScSZAl thin films deposited on Alloy 600 substrates using a 0.4 nm/s deposition rate.

The Raman spectra of ScSZ and ScSZAl thin films indicate mixed phases (Figure 4). Raman peaks are detected around 140, 262, 354, 382, 475, 540, 618, 726, 954, and 1000 cm^{-1}. The broad peak observed between 100 and 200 cm^{-1} consists of several peaks, indicating different ZrO_2 phases. The peaks at 354 and 382 cm^{-1} belong to the monoclinic phase [12,18,22], peaks detected at 147, 260, and 475 cm^{-1} belong to the tetragonal phase [15,21–24], and peaks detected around 540 cm^{-1} belong to the rhombohedral phase [12,24]. The broad peak expressed around 620 cm^{-1} consists of several peaks indicating the presence of cubic (c), tetragonal (t), and monoclinic (m) phases. On the other hand, Raman peaks near 620 cm^{-1} can be shifted due to a disordered oxygen sublattice after doping and co-doping by Sc and Al. Substitution of Zr^{4+} by Sc^{3+} results in the formation of high quantities of oxygen vacancies [37] and such a high defect concentration can lead to a violation of the selection rules and the appearance of additional modes that are forbidden for the cubic fluorite structure [38].

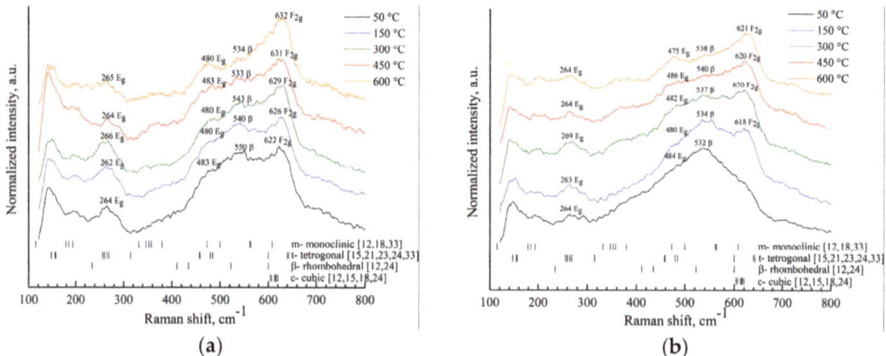

Figure 4. Raman spectra of (**a**) ScSZ and (**b**) ScSZAl thin films deposited on Alloy 600 substrates using a 0.4 nm/s deposition rate.

The Raman spectra of ScSZ and ScSZAl are of a similar shape; however, the intensities of peaks at 620 cm^{-1} are lower for the ScSZAl thin films (Figure 4). In the case of thin films with an Al dopant (Figure 4b), the cubic phase begins to form at a temperature higher than 50 °C and further increase of the temperature does not significantly influence the amount of the cubic phase. This means that Al slows down the cubic phase formation and stabilizes it at higher than 300 °C substrate temperatures. Quantitative calculations give the same substantiation (Table 2). The ratios of the cubic phase to the tetragonal and rhombohedral phases are from 42% to 53% for ScSZ and around 42% for ScSZAl. Calculations reveal that the amount of cubic phase increases by 10% for ScSZ and by 2% for ScSZAl with increasing substrate temperature. Finally, Raman analysis shows that thin films have a mixture of cubic (~44%), tetragonal (~18%), and rhombohedral (~38%) phases (Table 2). The obtained results demonstrat that a polymorphous transition from rhombohedral to cubic phase occurred in the formed thin films, indicating a typical behavior for the doped ZrO$_2$ [39]. According to the XRD spectra, the ZrO$_2$ rhombohedral phase is not significantly expressed and the tetragonal phase is not observed, while Raman spectra show large amounts of these phases, meaning that both phases could be located in the grain boundaries [15,39].

Table 2. The ratio of cubic, tetragonal, and rhombohedral phases in the ScSZ and ScSZAl thin films.

Substrate Temperature (°C)	Cubic	Tetragonal	Rhombohedral
ScSZ			
50	43%	15%	42%
150	42%	19%	39%
300	45%	18%	37%
450	46%	17%	37%
600	53%	17%	30%
ScAlSZ			
50	–	22%	78%
150	41%	13%	46%
300	42%	19%	39%
450	43%	20%	37%
600	43%	23%	34%

The average crystallite size dependence on the substrate temperature shows a nonlinear behavior (Figure 5). Crystallites grow larger (17.4–69.9 nm) with increasing substrate temperature (50–300 °C) during deposition. At 450 °C temperature, a sudden decrease of crystallite size occurs (~45–30 nm). This can be related to the changes in preferential orientation and to the changes in the ratio of phases. The preferential orientation changes from (200) at the temperatures up to 450 °C to (111) at high temperatures due to the higher diffusion energy of adatoms which allows them to move on the surface and become trapped on the high diffusivity surfaces.

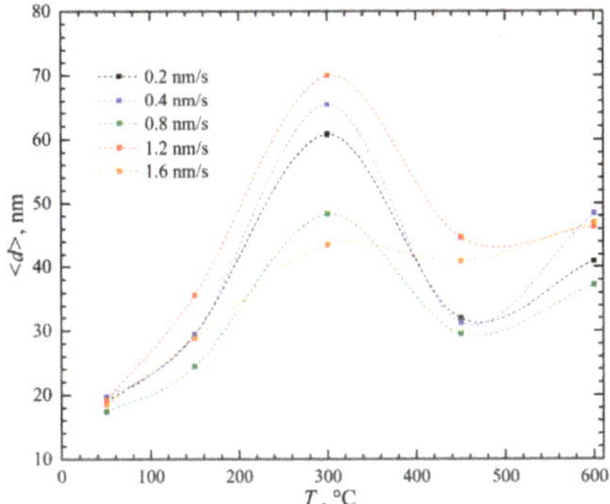

Figure 5. Crystallite size dependence on substrate temperature of ScSZAl thin films deposited on Alloy 600 substrates using different deposition rates.

Arrhenius plots show linear dependences (Figure 6a). No obvious breaking or bending points were observed. This means that no phase transitions occurred during the measurements. Moreover, ionic conductivity (600 °C measurement temperature) is observed to be related to the substrate temperature (Table 3). Ionic conductivity is higher for the thin films deposited on higher temperature substrates. The highest value of ionic conductivity of 4.2×10^{-3} S/cm (substrate temperature 600 °C and deposition rate 0.4 nm/s) is similar to other authors' results (~7×10^{-3} S/cm) [13,16,24]. Crystallite size has a similar dependence on temperature. Crystallites were observed to grow larger (17.4–69.9 nm) with increasing substrate temperature (50–300 °C) during deposition. At 450 °C, a sudden decrease of crystallite size occurs (~45–30 nm). This may be related to the changes in preferential orientation and to the changes in the ratio of phases. It is known that ionic conductivity is strongly related to crystallite size. According to the brick layer model, materials consisting of larger crystallites exhibit higher ionic conductivity because grain boundaries slow down oxygen ion diffusion [40–42].

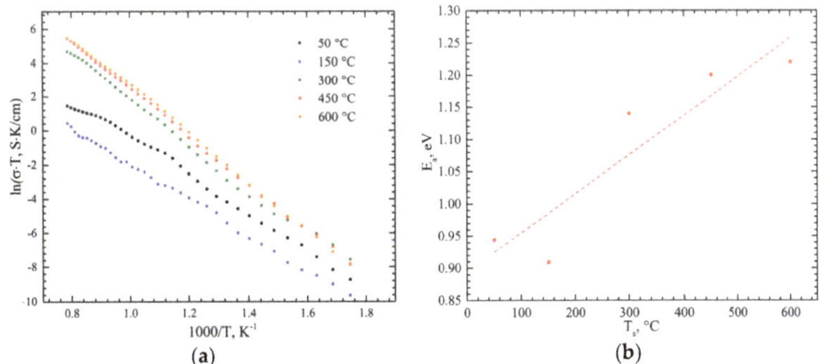

Figure 6. (a) Arrhenius plots of ScSZAl thin films deposited on Al_2O_3 substrates using a 0.4 nm/s deposition rate and (b) vacancy activation energy dependence on the substrate temperature of ScSZAl thin films deposited on Al_2O_3 substrates using a 0.4 nm/s deposition rate.

Table 3. Ionic conductivity (S/cm) of ScSZ and ScSZAl thin films deposited using different substrate temperature and a 0.4 nm/s deposition rate.

Thin Films	Substrate Temperature				
	50 °C	150 °C	300 °C	450 °C	600 °C
ScSZ	1.7×10^{-5}	2.5×10^{-5}	7.4×10^{-4}	3.6×10^{-3}	4.2×10^{-3}
ScAlSZ	4.1×10^{-5}	2.3×10^{-4}	1.1×10^{-3}	1.9×10^{-3}	2.4×10^{-3}

Co-doping of aluminum was observed to have a minor effect on the ionic conductivity of thin films, although conductivity is slightly lower for the ScSZAl thin films deposited on higher temperature substrates. The lower ionic conductivity is a result of a lower amount of cubic phase and a higher amount of tetragonal phase (Table 2).

It was noticed that vacancy activation energy increased from 0.91 to 1.22 eV using higher temperature substrates (Figure 6b). This increase also occurred due to the higher amount of tetragonal phase in those thin films (Table 3).

4. Conclusions

In this work, thin films of ScSZ and ScSZAl were deposited using electron beam vapor deposition, which allows the production of a dense and homogenous thin film. Structure and conductivity studies of the formed thin films were performed. It was found that the structure of the formed thin films does not repeat the structure of the initial evaporated material. Analysis of XRD patterns and Raman spectra of the initial evaporated powders of ScSZ and ScSZAl shows that they exhibit a polymorphism of ZrO_2 monoclinic, tetragonal (in Raman spectroscopy), and rhombohedral phases. Contrarily to the structure of evaporating material, XRD of ScSZ and ScSZAl thin films depict only a pure ZrO_2 face-centered cubic phase of a preferential orientation (200) at temperatures up to 450 °C, with a change in preferential orientation (100) at higher temperatures. In order to investigate the influence of an Al dopant, Rietveld analysis was performed, demonstrating that a pure cubic phase can be found in the ScSZAl thin films with equal probability as a mixture of cubic (87.8%) and rhombohedral (12.2%) phases. Raman spectra of ScSZ and ScSZAl thin films also indicate a polymorphism of ZrO_2 phases. Raman peaks detected around 140, 262, 354, 382, 475, 540, 618, 726, 954, and 1000 cm^{-1} indicate the presence of cubic (~44%), tetragonal (~18%) and rhombohedral (~38%) phases, showing a transition from a rhombohedral to a cubic phase. In ScSZAl thin films, co-doping with Al delays and slows down the formation of a cubic phase and stabilizes it at higher than 300 °C substrate temperatures. The ratios of the cubic phase to the tetragonal and rhombohedral phases are from 42% to 53% for ScSZ and around 42% for ScSZAl. It was found that the crystallite size depends on substrate temperature, demonstrating a nonlinear behavior. Crystallites grew larger (17.4–69.9 nm) with increasing substrate temperature (50–300 °C) and a sudden decrease of their size occurred (~45–30 nm) at a 450 °C deposition temperature due to the changes in preferential orientation and phase ratio. On the other hand, Arrhenius plots showed linear dependences and exhibited the highest value of ionic conductivity of 4.2×10^{-3} S/cm for thin films deposited on 600 °C temperature substrates using a 0.4 nm/s deposition rate. The increase in the amount of tetragonal phase in the formed thin films influences the vacancy activation energy which increases from 0.91 to 1.22 eV using higher temperature substrates.

Author Contributions: Conceptualization, M.S., G.L, and N.K.; methodology, K.B. and Ž.R.; formal analysis, D.V., M.S., and K.B.; investigation, D.V., N.K., M.S., and Ž.R.; writing—original draft preparation, M.S., N.K., K.B., Ž.R., D.V., and G.L.; writing—review and editing, M.S., N.K., K.B., Ž.R., D.V., and G.L.; visualization, N.K.; supervision, G.L.; project administration, K.B.; funding acquisition, G.L.

Funding: This research was funded by the European Regional Development Fund according to the supported activity "Research Projects Implemented by World-class Researcher Groups" under Measure No. 01.2.2-LMT-K-718.

Acknowledgments: Authors would like to express their gratitude for the following individuals for their expertise and contribution to the manuscript: Arvaidas Galdikas, Teresa Moskalioviene, Gediminas Kairaitis, and Matas Galdikas.

Conflicts of Interest: The authors declare no conflict of interest.

References

1. Kawada, T.; Mizusaki, J. Current electrolytes and catalysts. In *Handbook of Fuel Cells—Fundamentals, Technology and Applications*; Vielstich, W., Lamm, A., Eds.; John Wiley & Sons: Hoboken, NJ, USA, 2010.
2. Jung, W.C.; Hertz, J.L.; Tuller, H.L. Enhanced ionic conductivity and phase meta-stability of nano-sized thin film yttria-doped zirconia (YDZ). *Acta Mater.* **2009**, *57*, 1399–1404. [CrossRef]
3. Sternik, M.; Parlinski, K. Lattice vibrations in cubic, tetragonal, and monoclinic phases of ZrO_2. *J. Chem. Phys.* **2005**, *122*, 064707. [CrossRef] [PubMed]
4. Simeone, D.; Gosset, D.; Bechade, J.L.; Chevarier, A. Analysis of the monoclinic–tetragonal phase transition of zirconia under irradiation. *J. Nucl. Mater.* **2002**, *300*, 27–38. [CrossRef]
5. Liu, T.; Zhang, X.; Wang, X.; Yu, J.; Li, L. A review of zirconia-based solid electrolytes. *Ionics* **2016**, *22*, 2249–2262. [CrossRef]
6. Xue, Q.N.; Wang, L.G.; Huang, X.W.; Zhang, J.X.; Zhang, H. Influence of codoping on the conductivity of Sc-doped zirconia by first-principles calculations and experiments. *Mater. Des.* **2018**, *160*, 131–137. [CrossRef]
7. Agarkov, D.A.; Borik, M.A.; Bredikhin, S.I.; Burmistrov, I.N.; Eliseeva, G.M.; Kolotygin, V.A.; Kulebyakin, A.V.; Kuritsyna, I.E.; Lomonova, E.E.; Milovich, F.O.; et al. Structure and transport properties of zirconia crystals co-doped by scandia, ceria and yttria. *J. Mater.* **2019**. [CrossRef]
8. Taylor, M.A.; Argirusis, C.; Kilo, M.; Borchardt, G.; Luther, K.-D.; Assmus, W. Correlation between ionic radius and cation diffusion in stabilised zirconia. *Solid State Ion.* **2004**, *173*, 51–56. [CrossRef]
9. Mago, S.; Sharma, C.; Mehra, R.; Pandey, O.P.; Singh, K.L.; Singh, A.P. Preparation of YZT a mixed conductor by microwave processing: A different mechanism in the solid state thermochemical reaction. *Mater. Chem. Phys.* **2018**, *216*, 372–379. [CrossRef]
10. Mahato, N.; Gupta, A.; Balani, K. Doped zirconia and ceria-based electrolytes for solid oxide fuel cells: A review. *Nanomater. Energy* **2012**, *1*, 27–45. [CrossRef]
11. Arachi, Y.; Sakai, H.; Yamamoto, O.; Takeda, Y.; Imanishai, N. Electrical conductivity of the $ZrO_2–Ln_2O_3$ (Ln = lanthanides) system. *Solid State Ion.* **1999**, *121*, 133–139. [CrossRef]
12. Dasari, H.P.; Ahn, J.S.; Ahn, K.; Park, S.-Y.; Hong, J.; Kim, H.; Yoon, K.J.; Son, J.-W.; Lee, H.-W.; Lee, J.-H. Synthesis, sintering and conductivity behavior of ceria-doped Scandia-stabilized zirconia. *Solid State Ion.* **2014**, *263*, 103–109. [CrossRef]
13. Ishii, T. Structural phase transition and ionic conductivity in $0.88ZrO_2–(0.12 − x)Sc_2O_3–xAl_2O_3$. *Solid State Ion.* **1995**, *78*, 333–338. [CrossRef]
14. Sarat, S.; Sammes, N.; Smirnova, A. Bismuth oxide doped scandia-stabilized zirconia electrolyte for the intermediate temperature solid oxide fuel cells. *J. Power Sources* **2006**, *160*, 892–896. [CrossRef]
15. Wang, Z.; Cheng, M.; Bi, Z.; Dong, Y.; Zhang, H.; Zhang, J.; Feng, Z.; Li, C. Structure and impedance of ZrO_2 doped with Sc_2O_3 and CeO_2. *Mater. Lett.* **2005**, *59*, 2579–2582. [CrossRef]
16. Ota, Y.; Ikeda, M.; Sakuragi, S.; Iwama, Y.; Sonoyama, N.; Ikeda, S.; Hirano, A.; Imanishi, N.; Takeda, Y.; Yamamoto, O. Crystal structure and oxygen ion conductivity of Ga^{3+} Co-doped scandia-stabilized zirconia. *J. Electrochem. Soc.* **2010**, *157*, B1707–B1712. [CrossRef]
17. Yarmolenko, S.; Sankar, J.; Bernier, N.; Klimov, M.; Kapat, J.; Orlovskaya, N. Phase stability and sintering behavior of 10 mol % Sc_2O_3–1 mol % CeO_2–ZrO_2 ceramics. *J. Fuel Cell Sci. Technol.* **2009**, *6*, 21007–21008. [CrossRef]
18. Hirata, T.; Asari, E.; Kitajima, M. Infrared and Raman spectroscopic studies of ZrO_2 polymorphs doped with Y_2O_3 or CeO_2. *J. Solid State Chem.* **1994**, *110*, 201–207. [CrossRef]
19. Feinberg, A.; Perry, C.H. Structural disorder and phase transitions in $ZrO_2–Y_2O_3$ system. *J. Phys. Chem. Solids* **1981**, *42*, 513–518. [CrossRef]
20. Jawhari, T. Micro-Raman spectroscopy of the solid state: Applications to semiconductors and thin films. *Analusis* **2000**, *28*, 15–21. [CrossRef]

21. Nomura, K.; Mizutani, Y.; Kawai, M.; Nakamura, Y.; Yamamoto, O. Aging and Raman scattering study of scandia and yttria doped zirconia. *Solid State Ion.* **2000**, *132*, 235–239. [CrossRef]

22. Li, L.; Wang, W. Synthesis and characterization of monoclinic ZrO_2 nanorods by a novel and simple precursor thermal decomposition approach. *Solid State Commun.* **2003**, *127*, 639–643. [CrossRef]

23. Xue, Q.; Huang, X.; Zhang, H.; Xu, H.; Zhang, J.; Wang, L. Synthesis and characterization of high ionic conductivity ScSZ core/shell nanocomposites. *J. Rare Earths* **2017**, *35*, 567–573. [CrossRef]

24. Kumar, A.; Jaiswal, A.; Sanbui, M.; Omar, S. Scandia stabilized zirconia-ceria solid electrolyte (xSc1CeSZ, 5 < x < 11) for IT-SOFCs: Structure and conductivity studies. *Scr. Mater.* **2016**, *121*, 10–13. [CrossRef]

25. Jamil, Z.; Ruiz-Trejo, E.; Boldrin, P.; Brandon, N.P. Anode fabrication for solid oxide fuel cells: Electroless and electrodeposition of nickel and silver into doped ceria scaffolds. *Int. J. Hydrogen Energy* **2016**, *41*, 9627–9637. [CrossRef]

26. Shi, H.; Ran, R.; Shao, Z. Wet powder spraying fabrication and performance optimization of IT-SOFCs with thin-film ScSZ electrolyte. *Int. J. Hydrog. Energy* **2012**, *37*, 1125–1132. [CrossRef]

27. Ksapabutr, B.; Chalermkiti, T.; Wongkasemjit, S.; Panapoy, M. Fabrication of scandium stabilized zirconia thin film by electrostatic spray deposition technique for solid oxide fuel cell electrolyte. *Thin Solid Films* **2010**, *518*, 6518–6521. [CrossRef]

28. Pshyk, A.V.; Coy, E.; Kempiński, M.; Scheibe, B.; Jurga, S. Low-temperature growth of epitaxial Ti_2AlC MAX phase thin films by low-rate layer-by-layer PVD. *Mater. Res. Lett.* **2019**, *7*, 244–250. [CrossRef]

29. Xu, Z.; He, S.; He, L.; Mu, R.; Huang, G.; Cao, X. Novel thermal barrier coatings based on $La_2(Zr_{0.7}Ce_{0.3})_2O_7$/8YSZ double-ceramic-layer systems deposited by electron beam physical vapor deposition. *J. Alloy. Compd.* **2011**, *509*, 4273–4283. [CrossRef]

30. Lopatin, S.I.; Shugurov, S.M. Thermodynamic properties of the Lu_2O_3–ZrO_2 solid solutions by Knudsen effusion mass spectrometry at high temperature. *J. Chem. Thermodyn.* **2014**, *72*, 85–88. [CrossRef]

31. Jacobson, N.S. Thermodynamic properties of some metal oxide-zirconia systems (No. NASA-E-5060). In *NASA Technical Memorandum 102351*; NASA: Cleveland, OH, USA, 1989.

32. Fujimori, H.; Yashima, M.; Kakihana, M.; Yoshimura, M. Structural changes of Scandia-doped zirconia solid solutions: rietveld analysis and Raman scattering. *J. Am. Ceram. Soc.* **2005**, *81*, 2885–2893. [CrossRef]

33. Le Bail, A. The profile of a Bragg reflection for extracting intensities. In *Powder Diffraction: Theory and Practice*; Dinnebier, R.E., Billinge, S.J.L., Eds.; The Royal Society of Chemistry: London, UK, 2008; pp. 134–165.

34. Siu, G.G.; Stokes, M.J.; Liu, Y. Variation of fundamental and higher-order Raman spectra of ZrO_2 nanograins with annealing temperature. *Phys. Rev. B* **1999**, *59*, 3173–3179. [CrossRef]

35. Garvie, R.C. The occurrence of metastable tetragonal zirconia as a crystallite size effect. *J. Phys. Chem.* **1965**, *69*, 1238–1243. [CrossRef]

36. Petrov, I.; Barna, P.B.; Hultman, L.; Greene, J.E. Microstructural evolution during film growth. *J. Vac. Sci. Technol. A* **2003**, *21*, S117–S128. [CrossRef]

37. Marrocchelli, D.; Madden, P.A.; Norberg, S.T.; Hull, S. Structural disorder in doped zirconias, Part II: Vacancy ordering effects and the conductivity maximum. *Chem. Mater.* **2011**, *23*, 1365–1373. [CrossRef]

38. Kosacki, I.; Suzuki, T.; Anderson, H.U.; Colomban, P. Raman scattering and lattice defects in nanocrystalline CeO_2 thin films. *Solid State Ion.* **2002**, *149*, 99–105. [CrossRef]

39. Spirin, A.; Ivanov, V.; Nikonov, A.; Lipilin, A.; Paranin, S.; Khrustov, V.; Spirina, A. Scandia-stabilized zirconia doped with yttria: Synthesis, properties, and ageing behavior. *Solid State Ion.* **2012**, *225*, 448–452. [CrossRef]

40. Maier, J. On the conductivity of polycrystalline materials. *Berichte der Bunsengesellschaft für Physikalische Chemie* **1986**, *90*, 26–33. [CrossRef]

41. Tschöpe, A.; Sommer, E.; Birringer, R. Grain size-dependent electrical conductivity of polycrystalline cerium oxide: I. Experiments. *Solid State Ion.* **2001**, *139*, 255–265. [CrossRef]

42. Tao, J.; Dong, A.; Wang, J. The influence of microstructure and grain boundary on the electrical properties of scandia stabilized zirconia. *Mater. Trans.* **2013**, *54*, 825–832. [CrossRef]

Article

Fabrication of a Conjugated Fluoropolymer Film Using One-Step iCVD Process and its Mechanical Durability

Hyo Seong Lee [1], Hayeong Kim [2], Jeong Heon Lee [1] and Jae B. Kwak [1,*]

[1] Department of Mechanical System and Automotive Engineering, Chosun University, 309 Pilmun-daero, Dong-gu, Gwangju 61452, Korea
[2] Global Technology Center, Samsung Electronics Co., Ltd., 129 Samsung-ro, Yeongtong-gu, Suwon-si 16677, Korea
* Correspondence: jaekwak@chosun.ac.kr; Tel.: +82-62-230-7095

Received: 12 June 2019; Accepted: 4 July 2019; Published: 8 July 2019

Abstract: Most superhydrophobic surface fabrication techniques involve precise manufacturing process. We suggest initiated chemical vapor deposition (iCVD) as a novel CVD method to fabricate sufficiently durable superhydrophobic coating layers. The proposed method proceeds with the coating process at mild temperature (40 °C) with no need of pretreatment of the substrate surface; the pressure and temperature are optimized as process parameters. To obtain a durable superhydrophobic film, two polymeric layers are conjugated in a sequential deposition process. Specifically, 1,3,5,7-tetravinyl-1,3,5,7-tetramethylcyclotetrasiloxane (V4D4) monomer is introduced to form an organosilicon layer (pV4D4) followed by fluoropolymer formation by introducing 1H, 1H, 2H, 2H-Perfluorodecyl methacrylate (PFDMA). There is a high probability of covalent bond formation at the interface between the two layers. Accordingly, the mechanical durability of the conjugated fluoropolymer film (pV4D4-PFDMA) is reinforced because of cross-linking. The superhydrophobic coating on soft substrates, such as tissue paper and cotton fabric, was successfully demonstrated, and its durability was assessed against the mechanical stress such as tensile loading and abrasion. The results from both tests confirm the improvement of mechanical durability of the obtained film.

Keywords: initiated chemical vapor deposition (iCVD); superhydrophobic; fluoropolymer

1. Introduction

Superhydrophobic surfaces attracts considerable research attention as they are used in many industrial applications, including water repellents, antifouling, and self-cleaning surfaces [1–12]. There are many methods and techniques to produce superhydrophobic surfaces such as metal etching [2,3], sol-gel processing [4], dip-spraying [5], electro-spray and spinning [6,11], combining solution and phase inversion process [7,8], solution method [9], ultrasound-assisted method [10], and plasma-based method [12]. However, most of these techniques are only suitable for nanostructures with fine surface roughness and are further limited to specific substrates such as metals. A hierarchical structure is always better for a superhydrophobic surface because it can reduce the contact area between the water and the surface. This surface can be obtained using a combination of the topographical properties of the surface texture and the chemical properties of low surface energy. To demonstrate this, Zhou et al. have synthesized nanostructured ZnO film with super-hydrophobicity using a chemical vapor deposition (CVD) method [13].

In comparison with other methods, the CVD process can produce large area of homogeneous thin films. However, it is not readily available for flexible substrates such as cotton fabrics and papers because it requires the use of high temperature and high vacuum pressure. For commercial purposes, a superhydrophobic surface needs to be mechanically and chemically durable. For instance, Zhou et al. fabricated a durable superhydrophobic polyester fabric with fluoroalkyl silane-modified silicone rubber, which is a nanoparticle composite, using the dip coating method [14]. Yan et al. also fabricated hydrophobic and hydrophilic surface of cotton fabric using a plasma-induced graft method [15]. Recently, Heydari Gharahcheshmeh and Gleason reported the fabrication of an antifouling surface on conductive polymer film via the oxidative chemical vapor deposition (oCVD) process that improves signal-to-noise ratio of the neural recording electrodes against blood or tissue contamination [16]. In this study, we use an iCVD process to generate fluoropolymer film for super hydrophobicity because this is relatively simple process in comparison with the previous methods. Unlike the conventional CVD process, the iCVD is achievable at a low temperature and vacuum pressure and shows excellent step coverage at high aspect ratio structure [17].

Figure 1 shows the iCVD process. The process begins with canisters containing an initiator (I_2) and one or more monomers (M), which are the building blocks of the desired polymer coating. These materials are vaporized, either by heating or reducing the air pressure, and are introduced simultaneously into a vacuum chamber with the substrate to be coated. Once vaporized, the initiator molecules are thermally decomposed upon the contact with a hot filament to become radicals (I*). The radicals activate the vaporized monomer to link in chains that form polymers on the surface of the substrate kept at mild temperature (25–40 °C). This is a one-step, solvent-free process to grow polymer films uniformly onto complex substrates structures, regardless of the substrate material. The functional performance of the polymer thin film is attributed to the properties of the monomers used [18].

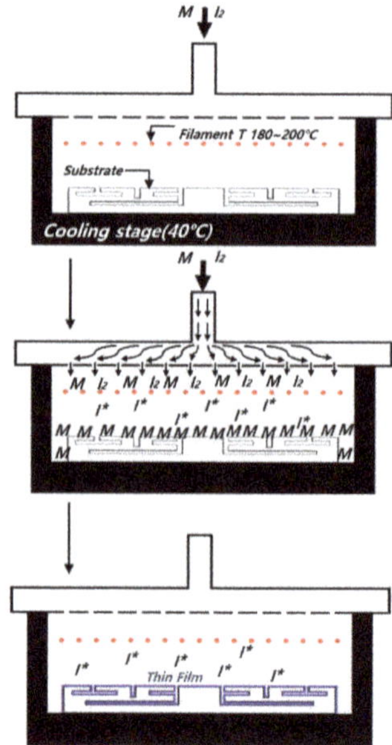

Figure 1. A schematic of iCVD process.

To obtain the fluoropolymer film with the iCVD process, tert-butyl peroxide (TBPO) and 1H, 1H, 2H, 2H-perfluorodecyl acrylate (PFDA) are typically used as the initiator and the monomer, respectively [19–22]. In particular, 1H, 1H, 2H, 2H-perfluorodecyl methacrylate (PFDMA) is of great interest because of its exceptionally low surface energy (3.5 mN·m^{-1}), similar to PFDA and more efficient commercialization possibilities [23,24]. These monomers are synthesized in a one-step iCVD process and show more than 150° of water contact angle. Previously, mechanical and chemical robustness of the iCVD superhydrophobic coating had not been studied. Therefore, 1,3,5,7-tetravinyl-1,3,5,7-tetramethylcyclotetrasiloxane (V4D4) was introduced as a cross linker, which improves mechanical strength. Finally, a conjugated polymer film was obtained with a successive deposition of an organosilicon polymer, pV4D4 (poly-tetramethylcyclotetrasiloxane) and a fluoropolymer, pPFDMA (poly-perfluorodecyl methacrylate) using the customized iCVD reactor. Because pPFDMA is strongly bound on the pV4D4 layer, the durability of the conjugated film (pV4D4-PFDMA) improves due to increasing adhesion strength between the substrate and the fluoropolymer film. In this study, only pPFDMA, pV4D4-PFDA, and pV4D4-PFDMA were synthesized with iCVD, after which the durability of the each film was evaluated using the rubbing test. In addition, infiltration capability was examined after coating the folded copper sheet.

2. Experiment

2.1. iCVD Process Basics

The iCVD reaction mechanism is initiated by pyrolysis of the initiator to radicalize non-covalent electrons to bond with the vinyl group of the monomer and to grow the polymer thin film through the sequential chain bonding [25,26]. Notably, the initiator TBPO is radicalized through the pyrolysis and polymerized with the vinyl groups of monomers such as V4D4 and PFDMA, resulting in thin film growth. As shown in Equations (1)–(3), the parameter of iCVD process is defined as an optimal value of P_M/P_{Sat}. The P_M and P_{Sat} values are derived from the Antoine Equations (1) and (2), where P_M is the ratio of the incoming gas amount controlled by the vaporization temperature of the monomer and the initiator, and the P_{Sat} value is determined by monomer's characterization and the temperature of the substrate ($T_{substrate}$). The optimized P_M/P_{Sat} is between 0 and 1, which is the rate of polymerization. When P_M/P_{Sat} is close to 0, the growth rate is very slow. When it is close to 1, condensation occurs instead of polymerization. Also, the duration of the deposition affects the thickness of the polymerized film. Optimized P_M/P_{Sat} varies depending on monomers.

$$P_M = P_{Chamber} \times \frac{F_M}{F_M + F_I} \tag{1}$$

$$\log P_{Sat} = A - \frac{B}{T_{substrate}} \tag{2}$$

$$0 < \frac{P_M}{P_{Sat}} < 1 \tag{3}$$

where, F_M (sccm) is a gaseous monomer input flow, F_I (sccm) is a gaseous initiator input flow, P_M (mTorr) is the partial pressure of the monomer in the chamber, P_{Sat} (mTorr) is the saturated vapor pressure at the $T_{Substrate}$, $P_{Chamber}$ (mTorr) is the total pressure of the chamber, and $T_{Substrate}$ (°C) is the temperature of substrate.

The customized iCVD reactor system uses canisters for monomers and the initiator. The canisters are heated to efficiently vaporize the materials into the vacuum chamber. Also, there are filaments arrayed in the vacuum chamber to instantly apply the elevated temperature for decomposing the initiator into radicals.

2.2. Fabrication of Superhydrophobic Film and its Characteristics

Table 1 shows the monomers and the initiator examined in this study. The deposition of pV4D4-PFDMA was performed in the iCVD reactor. V4D4 and PFDMA monomers were heated to 55 and 75 °C, respectively, and TBPO was heated to 35 °C through each canister. Each flow rate of the vapor was controlled by needle valves, and the flow was monitored by a pressure gauge installed at the inlet of the reactor. In this study, for all monomers and initiator, the vapor saturation ratio ($F_M{:}F_I$) was maintained to be 1:1 as empirically optimized P_M/P_{sat}. First, V4D4 and TBPO were delivered into the reactor at 0.6 sccm while maintaining 270 mTorr vacuum pressure in the reactor. The substrate and the filaments temperature were held at 40 and 180 °C, respectively. Then, a homopolymer of pV4D4 was grown from the surface of substrate. Once the desired thickness of pV4D4 was obtained, the next vaporized PFDMA was introduced into the reactor together with V4D4 for 2–3 min. This period ensures strong adhesion at the film interface between the top pV4D4 layer and the bottom pPFDMA layer. Then, the flow of V4D4 was stopped, and the deposition of pPFDMA continued to the top layer. With PFDMA, only the vacuum pressure in the reactor was changed to 100 mTorr to maintain the optimal P_M value; other temperature settings remained the same. As a result, a conjugated fluoropolymer film was obtained as a superhydrophobic surface, as shown in Figure 2a. The film was sequentially stacked pV4D4 and pV4D4-PFDMA on a silicon wafer substrate. Figure 2b shows the morphology of the fluoropolymer film; one can see about 5 µm size fluorocarbon structures tangled from the top view. To confirm the super-hydrophobicity of film, the contact angle was measured with 50 µL of deionized water droplets using a contact angle meter (SmartDrop, FEMTOBIOMED, Seongnam, Korea). As a result, the water contact angle of 150.1° ± 3.6° was obtained while the sliding angle was only 8.5°. This means that the pV4D4-PFDMA film exhibits a highly non-stick super-hydrophobic surface.

Table 1. Chemical structures and functions of the initiator and the monomers.

Structure	Name (Abbreviation)	Function	Chemical Formula
	tert-butyl peroxide (TBPO)	Initiator	$C_8H_{18}O_2$
	1H,1H,2H,2H-Perfluorodecyl acrylate (PFDA)	Super-hydrophobic	$C_{13}H_7F_{17}O_2$
	1H,1H,2H,2H-Perfluorodecyl methacrylate (PFDMA)	Super-hydrophobic	$C_{14}H_9F_{17}O_2$
	1,3,5,7-tetravinyl-1,3,5,7-tetramethylcyclotetrasiloxane (V4D4)	Cross-linker	$C_{12}H_{24}O_4Si_4$

Because the temperature of the substrate was kept at 40 °C, relatively flexible substrates were considered such as paper and fabric. Accordingly, we choose tissue paper and cotton fabric as substrates and proceeded with the pV4D4-PFDMA coating process in the iCVD reactor. At this time, the total thickness of the film was 200 nm so there were no significant changes in dimensions of the fiber structures. Both tissue and fabric are known to easily get wet. After coating, however, blue ink was dropped on the surface, and excellent repellency to the droplet was observed, as shown in Figure 3a,b. In addition, we performed SEM analysis (FE-SEM, S-4800, Hitachi, Tokyo, Japan), on both substrates before and after coating as shown in Figure 3b–f, respectively. No noticeable changes were observed.

Another advantage of the iCVD process is that it shows ultra-step coverage on high aspect ratio of up to 1:40 for the opened area [27]. Therefore, we made a quantitative analysis of the capability of coating with shadow area. A manifolded copper sheet was subjected to the iCVD process and coated with 200 nm of the pV4D4-PFDMA, as seen in Figure 4a. The thickness and the width of the copper sheet were 0.02 and 25 mm, respectively, and it was firmly folded seven times, as seen in Figure 4b. Next, the gap between each surface was measured about 0.1 mm. The pure copper sheet had a hydrophilic surface, as seen in Figure 4c; therefore, water droplets on various places of the unfolded sheet indicated the successful coating, as seen in Figure 4d. Additionally, Fourier transform infrared spectra (Nicolet 6700 FT-IR Spectrometer, Thermo Scientific, Waltham, MA, USA) analysis has proven an infiltration capability, as shown in Figure 4e. Although the peaks are not so obvious due to an extremely high aspect ratio for the shadow area, the result verifies the presence of the fluorocarbon functional group.

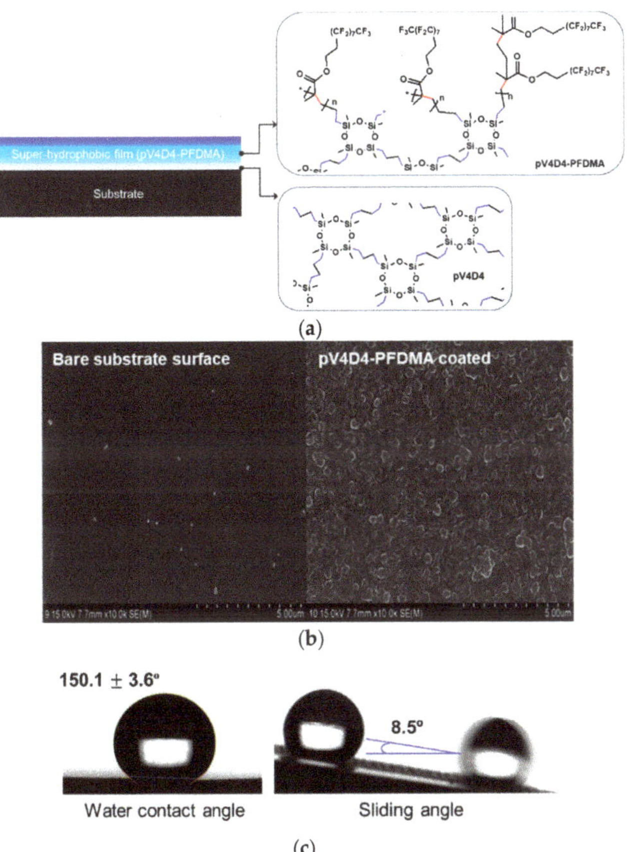

Figure 2. (**a**) A schematic of the fluoropolymer (pV4D4-PFDMA) coating; (**b**) surface morphology using scanning electron microscope (FE-SEM, S-4800, Hitachi, Tokyo, Japan) imaging before and after coating on wafer substrate; (**c**) static and dynamic water contact angles of the pV4D4-PFDMA surface.

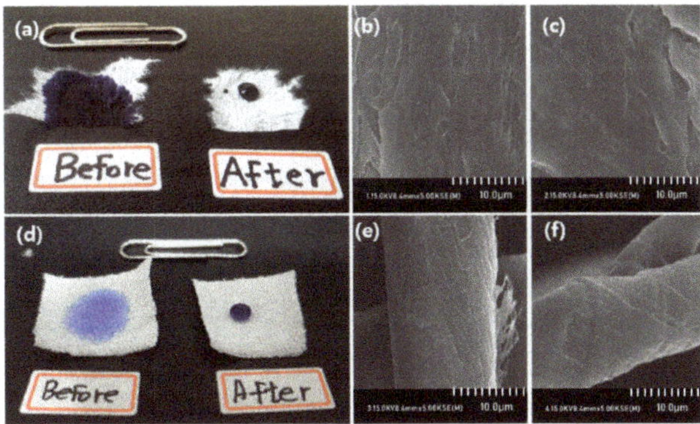

Figure 3. (**a**) Blue ink droplet on tissue paper before and after pV4D4-PFDMA coating; (**b**) SEM image of the tissue paper; (**c**) SEM image of the tissue structure after coating; (**d**) blue ink droplet on cotton fabric before and after pV4D4-PFDMA coating; (**e**) SEM image of the cotton fabric; and (**f**) SEM image of the cotton fabric after coating.

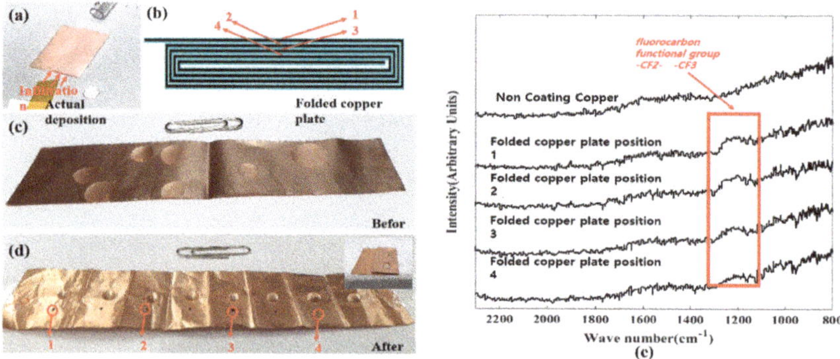

Figure 4. (**a**) A manifolded copper sheet in the iCVD reactor; (**b**) schematic of the folded copper sheet and FT-IR Spectrometer measurement locations; (**c**) water droplets on bare copper sheet; (**d**) water droplets on unfolded copper sheet after coating; and (**e**) FT-IR analysis before and after coating.

2.3. Examination of Durability

The development of a thin film that is resistant to various mechanical stresses is the key for successful commercialization of the superhydrophobic surfaces. The proposed film was obtained using the iCVD process, with a mechanically robust organosilicon polymer and a fluoropolymer layer. The two layers were covalently adhered to form a conjugated polymer structure. We evaluated the durability of the proposed coating against tensile and abrasion tests.

To compare the mechanical durability against deformation under external force between pPFDMA and pV4D4-PFDMA, we examined optical microscopic images of the cracks developed when the tensile load is applied. The pPFDMA (500 nm) and pV4D4 (100 nm)-PFDMA (500 nm) were deposited on the elastomer film (50 μm thickness) and subjected to tensile tests using Instron tester (5960 Series, INSTRON, Chicago, IL, USA) according to ASTM standard [28], as shown in Figure 5a. Tensile tests for each thin film were proceeded with the strain rate of 2.5 mm/min, and the obtained stress-strain plots were compared with the bare elastomer film, pPFDMA coated, and pV4D4-PFDMA coated, as seen in Figure 5d. According to the result, the pPFDMA coating shows the tensile strength was similar to that of the bare elastomer film; surface cracking was observed when the strain reached about 60%, as seen in Figure 5b,c. However, for pV4D4-PFDMA coating, crack occurrence was observed at 110% of strain, indicating an improvement of the adhesion strength between the elastomer film and the fluoropolymer film, resulting in eventual enhancement of the global tensile strength.

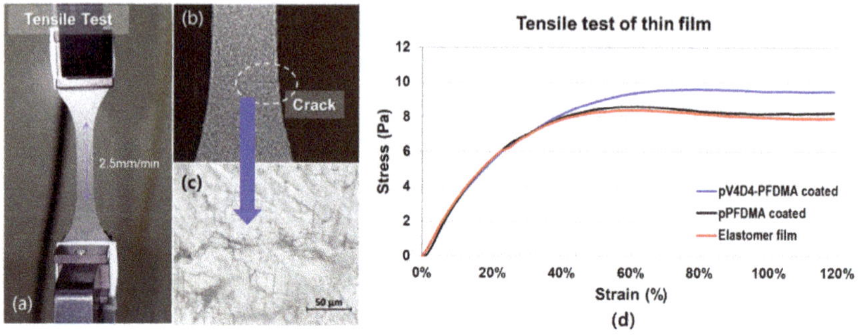

Figure 5. Tensile test of the elastomer film with superhydrophobic coating: (**a**) tensile test set-up; (**b**) surface cracks occurred during tensile loading; (**c**) microscopic image of the surface cracks; and (**d**) stress-strain curves.

The combined pV4D4 and pPFDMA film showed better stability and the super-hydrophobicity because of the covalent bonding between the interlayers. In this structure, benzene ring of the V4D4 acts as a cross linker, protecting the conjugated pV4D4–PFDMA from external stresses that normally occur in daily life. Accordingly, we evaluated the resistance in the abrasion test (CT-RB Series, CORETECH, Uiwang, Korea) and compared pPFDMA (500 nm) only, pV4D4 (100 nm)-PFDA (500 nm), and pV4D4 (100 nm)-PFDMA (500 nm). For the abrasion test, these films were deposited onto the SUS304 plate; the comparison was made using a contact angle meter. Figure 6a shows the abrasion test set-up. The coated sample was placed and firmly fixed in the sample holder. Then, the cotton-covered ball-shape tip was released to contact the top surface of the coated sample. 1 kgf weight was loaded on the tip, which was repeatedly rubbed on the surface left- and right-hand side throughout 3000 cycles. The contact angle was measured at every five cycles in the beginning and intermittently after hundreds of cycles. In abrasion test, the coated layer appeared to be scratched on the surface, as seen in Figure 6c and contact angle was gradually decreased. Figure 6d shows the comparison of the contact angle change for each film. According to the results, pPFDMA only showed the lowest abrasion resistance indicating drastic reduction of the contact angle within less than 10 cycles. However, the other two fluoropolymer films with pV4D4 conjugation withstood 200 cycles of rubbing and retained contact angle of 120°, which is considered as sufficiently high hydrophobicity. The proposed pV4D4-PFDMA showed the highest abrasion resistance, retaining 120° of the contact angle after 3000 cycles.

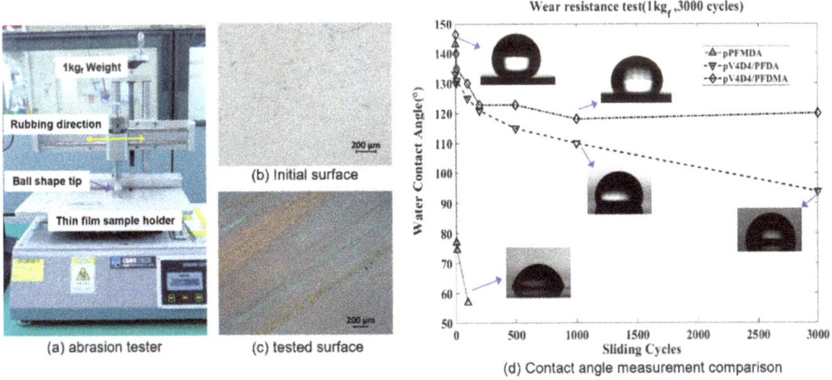

Figure 6. (**a**) Abrasion test set-up; (**b**) microscopic image of the initial surface; (**c**) microscopic image of the tested surface; and (**d**) contact angle measurements.

3. Conclusions

A fluoropolymer fabricated using the iCVD process has great advantages such as cost efficiency and high functionality when compared with other surface modification solutions to obtain super hydrophobicity. In this study, we examined in detail parametric optimization of the iCVD process and successfully fabricated a new super-hydrophobic film conjugated with V4D4 and PFDMA. The proposed superhydrophobic film was applied to tissue paper and cotton fabric and demonstrated great liquid repellency. In addition, high infiltration capability of the iCVD process was discussed using manifolded copper sheet. These results provide enough insight for industrial applications in which superhydrophobic surfaces are needed.

A conjugated film (pV4D4-PFDMA) was achieved by adding the pV4D4 layer before the introduction of the pPFDMA and showed exceptional stability and durability. The pV4D4 significantly enhances mechanical stability of the pPFDMA as it allowed for both monomers to flow into the reactor. Therefore, the fluoropolymer was reinforced by binding the organosilicon layer. We evaluated the mechanical and chemical robustness of the proposed film. First, the tensile test was performed using the deposition on the elastomer film; it showed improvement both in terms of the tensile strength and delay in surface cracking. Second, the mechanical abrasion test was performed and the proposed film showed better rubbing resistance when compared with other films.

The deposition process is applicable to various types and complex shapes of the substrates without the need of surface pretreatment; is also allows for improved adhesion between the coated film and the substrate. We found that the proposed superhydrophobic film obtained in the iCVD process provides an industrial grade of low surface energy with sufficient durability against various mechanical stresses. In addition, optimization of iCVD process can be further studied to enhance mechanical properties.

Author Contributions: Conceptualization, J.B.K.; Data Curation, H.K.; Formal Analysis, J.H.L.; Investigation, H.S.L. and J.B.K; Methodology, H.K.; Writing—Original Draft, H.S.L.; Writing—Review and Editing, J.B.K.

Funding: This research was funded by National Research Foundation of Korea (NRF) grant funded by the Korea government (MSIT) (No. NRF-2018R1C1B5045726) and Korea Institute for Advancement of Technology (KIAT) grant funded by the Korea Government (MOTIE) (P0002092, The Competency Development Program for Industry Specialist).

Conflicts of Interest: The authors declare no conflict of interest.

References

1. Wu, D.; Guo, Z. Robust and muti-repaired superhydrophobic surfaces via one-step method on copper and aluminum alloys. *Mater. Lett.* **2018**, *213*, 290–293. [CrossRef]
2. Latthe, S.S.; Sudhagar, P.; Devadoss, A.; Kumar, A.M.; Liu, S.; Terashima, C.; Nakata, K.; Fujishima, A. A mechanically bendable superhydrophobic steel surface with self-cleaning and corrosion-resistant properties. *J. Mater. Chem. A* **2015**, *3*, 14263–14271. [CrossRef]
3. He, J.; Wu, M.; Zhang, R.; Liu, J.; Deng, Y.; Guo, Z. A one-step hot-embossing process for fabricating a channel with superhydrophobic inner walls. *J. Manuf. Process.* **2018**, *36*, 351–359. [CrossRef]
4. Liu, S.; Liu, X.; Latthe, S.S.; Gao, L.; An, S.; Yoon, S.S.; Liu, B.; Xing, R. Self-cleaning transparent superhydrophobic coatings through simple sol-gel processing of fluoroalkylsilane. *Appl. Surf. Sci.* **2015**, *351*, 897–903. [CrossRef]
5. Latthe, S.S.; Sutar, R.S.; Kodag, V.S.; Bhosale, A.K.; Kumar, A.M.; Sadasivuni, K.K.; Xing, X.; Liu, S. Self–cleaning superhydrophobic coatings: Potential industrial applications. *Prog. Org. Coat.* **2019**, *128*, 52–58. [CrossRef]
6. Li, Y.; Men, X.; Zhu, X.; Ge, B.; Chu, F.; Zhang, Z. One-step spraying to fabricate nonfluorinated superhydrophobic coatings with high transparency. *J. Mater. Sci.* **2016**, *51*, 2411–2419. [CrossRef]
7. Fang, Z.; Li, H.; Xia, M.; Liu, Y.; Zhang, Y.; Li, Y.; Jiang, X.; He, P. A facile method to fabricate super-hydrophobic surface with water evaporation-induced phase inversion of stearic acid. *Mater. Lett.* **2018**, *223*, 124–127. [CrossRef]
8. Zhang, J.; Xu, Z.; Mai, W.; Min, C.; Zhou, B.; Shan, M.; Li, Y.; Yang, C.; Wang, Z.; Qian, X. Improved hydrophilicity, permeability, antifouling and mechanical performance of PVDF composite ultrafiltration membranes tailored by oxidized low-dimensional carbon nanomaterials. *J. Mater. Chem. A* **2013**, *1*, 3101–3111. [CrossRef]
9. Zhang, L.; Zuo, W.; Li, T. Controlled synthesis of bifunctional $PbWO_4$ dendrites via a facile solution method at room temperature: Photoluminescence and superhydrophobic property. *Mater. Sci. Semicond. Process.* **2015**, *39*, 188–191. [CrossRef]
10. Perkas, N.; Amirian, G.; Girshevitz, O.; Gedanken, A. Hydrophobic coating of GaAs surfaces with nanostructured ZnO. *Mater. Lett.* **2016**, *175*, 101–105. [CrossRef]
11. Rivero, P.J.; Iribarren, A.; Larumbe, S.; Palacio, J.F.; Rodríguez, R. A comparative study of multifunctional coatings based on electrospun fibers with incorporated ZnO nanoparticles. *Coatings* **2019**, *9*, 367. [CrossRef]
12. Ellinas, K.; Tserepi, A.; Gogolides, E. Superhydrophobic fabrics with mechanical durability prepared by a two-step plasma processing method. *Coatings* **2018**, *8*, 351. [CrossRef]
13. Zhou, M.; Feng, C.; Wu, C.; Ma, W.; Cai, L. Superhydrophobic multi-scale ZnO nanostructures fabricated by chemical vapor deposition method. *J. Nanosci. Nanotechnol.* **2009**, *9*, 4211–4214. [CrossRef] [PubMed]

14. Zhou, H.; Wang, H.; Niu, H.; Gestos, A.; Wang, X.; Lin, T. Fluoroalkyl silane modified silicone rubber/nanoparticle composite: A super durable, robust superhydrophobic fabric coating. *Adv. Mater.* **2012**, *24*, 2409–2412. [CrossRef]
15. Yan, X.; Li, J.; Yi, L. Fabrication of pH-responsive hydrophilic/hydrophobic Janus cotton fabric via plasma-induced graft polymerization. *Mater. Lett.* **2017**, *208*, 46–49. [CrossRef]
16. Heydari Gharahcheshmeh, M.; Gleason, K.K. Device fabrication based on oxidative chemical vapor deposition (oCVD) synthesis of conducting polymers and related conjugated organic materials. *Adv. Mater. Interfaces* **2019**, *6*, 1801564. [CrossRef]
17. Baxamusa, S.H.; Gleason, K.K. Thin polymer films with high step coverage in microtrenches by initiated CVD. *Chem. Vap. Depos.* **2008**, *14*, 313–318. [CrossRef]
18. Martin, T.P.; Lau, K.K.; Chan, K.; Mao, Y.; Gupta, M.; O'Shaughnessy, W.S.; Gleason, K.K. Initiated chemical vapor deposition (iCVD) of polymeric nanocoatings. *Surf. Coat. Technol.* **2007**, *201*, 9400–9405. [CrossRef]
19. Mao, Y.; Gleason, K.K. Positive-tone nanopatterning of chemical vapor deposited polyacrylic thin films. *Langmuir* **2006**, *22*, 1795–1799. [CrossRef]
20. Gupta, M.; Gleason, K.K. Initiated chemical vapor deposition of poly(1H, 1H, 2H, 2H-perfluorodecyl acrylate) thin films. *Langmuir* **2006**, *22*, 10047–10052. [CrossRef]
21. Ma, M.; Gupta, M.; Li, Z.; Zhai, L.; Gleason, K.K.; Cohen, R.E.; Rubner, M.F.; Rutledge, G.C. Decorated electrospun fibers exhibiting superhydrophobicity. *Adv. Mater.* **2007**, *19*, 255–259. [CrossRef]
22. Ma, M.; Mao, Y.; Gupta, M.; Gleason, K.K.; Rutledge, G.C. Superhydrophobic fabrics produced by electrospinning and chemical vapor deposition. *Macromolecules* **2005**, *38*, 9742–9748. [CrossRef]
23. Kim, J.H.; Park, S.K.; Lee, J.W.; Yoo, B.; Lee, J.K.; Kim, Y.H. Hydrophobic perfluoropolymer thin-film encapsulation for enhanced stability of inverted polymer solar cells. *J. Korean Phys. Soc.* **2014**, *65*, 1448–1452. [CrossRef]
24. Tsibouklis, J.; Graham, P.; Eaton, P.J.; Smith, J.R.; Nevell, T.G.; Smart, J.D.; Ewen, R.J. Poly(perfluoroalkyl methacrylate) film structures: surface organization phenomena, surface energy determinations, and force of adhesion measurements. *Macromolecules* **2000**, *33*, 8460–8465. [CrossRef]
25. Gleason, K.K. *CVD Polymer: Fabrication of Organic Surfaces and Devices*; WILET-VCH: Weinheim, Germany, 2015.
26. Moni, P.; Al-Obeidi, A.; Gleason, K.K. Vapor deposition routes to conformal polymer thin films. *Beilstein J. Nanotechnol.* **2017**, *8*, 723–735. [CrossRef] [PubMed]
27. Chan, K.; Gleason, K.K. A mechanistic study of initiated chemical vapor deposition of polymers: analyses of deposition rate and molecular weight. *Macromolecules* **2006**, *39*, 3890–3894. [CrossRef]
28. *ASTM D882-18: Standard Test Method for Tensile Properties of Thin Plastic Sheeting*; ASTM International: West Conshohocken, PA, USA, 2018.

Article

Simple Non-Destructive Method of Ultrathin Film Material Properties and Generated Internal Stress Determination Using Microcantilevers Immersed in Air

Ivo Stachiv [1,2,3,*] and Lifeng Gan [1]

[1] School of Sciences, Harbin Institute of Technology, Shenzhen, Shenzhen 518055, China
[2] Institute of Physics, Czech Academy of Sciences, 18221 Prague, Czech Republic
[3] Drážní revize s.r.o., Místecká 1120/103, 70300 Ostrava–Vitkovice, Czech Republic
* Correspondence: stachiv@fzu.cz; Tel.: +420-586532427

Received: 5 June 2019; Accepted: 27 July 2019; Published: 1 August 2019

Abstract: Recent progress in nanotechnology has enabled to design the advanced functional micro-/nanostructures utilizing the unique properties of ultrathin films. To ensure these structures can reach the expected functionality, it is necessary to know the density, generated internal stress and the material properties of prepared films. Since these films have thicknesses of several tens of nm, their material properties, including density, significantly deviate from the known bulk values. As such, determination of ultrathin film material properties requires usage of highly sophisticated devices that are often expensive, difficult to operate, and time consuming. Here, we demonstrate the extraordinary capability of a microcantilever commonly used in a conventional atomic force microscope to simultaneously measure multiple material properties and internal stress of ultrathin films. This procedure is based on detecting changes in the static deflection, flexural and torsional resonant frequencies, and the corresponding quality factors of the microcantilever vibrating in air before and after film deposition. In contrast to a microcantilever in vacuum, where the quality factor depends on the combination of multiple different mechanical energy losses, in air the quality factor is dominated just by known air damping, which can be precisely controlled by changing the air pressure. Easily accessible expressions required to calculate the ultrathin film density, the Poisson's ratio, and the Young's and shear moduli from measured changes in the microcantilever resonant frequencies, and quality factors are derived. We also show that the impact of uncertainties on determined material properties is only minor. The validity and potential of the present procedure in material testing is demonstrated by (i) extracting the Young's modulus of atomic-layer-deposited TiO_2 films coated on a SU-8 microcantilever from observed changes in frequency response and without requirement of knowing the film density, and (ii) comparing the shear modulus and density of Si_3N_4 films coated on the silicon microcantilever obtained numerically and by present method.

Keywords: thin film; atomic layer deposition; nanomechanics; Young's modulus; shear modulus; resonant frequency; Q-factor; microcantilevers; internal stress

---

1. Introduction

Functional micro-/nanostructures made of substrate and one or multiple ultrathin films are widely used in applications like photovoltaics [1], micro-electronics [2], optics [3,4], tunable resonators [5,6], and various sensors [7–12]. Preparation of these structures involves repeated usage of multiple fabrication processes such as deposition, lithography, etching, and cleaning. In order to prevent the mechanical failure or to guarantee that the structures would reach the desired operating conditions, the material properties of prepared ultrathin films must be known. During film deposition the internal stress that often originates from a coefficient of thermal expansion mismatch can be generated [13]. For films of thicknesses ranging from hundreds of nm to about tens of μm, the nanoindentation [14], bulge test [15], and the resonant methods [16] are the most common techniques used to determine the film material properties. Noticing that for micro-sized samples the density is estimated based on the known bulk values. However, once film thickness shrinks from micro- to nanoscale (i.e., tens of nm), the density of particularly polymer, organic, composite, or porous films can start to deviate from the bulk values (e.g., changes in deposition parameters affect notably prepared film density) [17,18]. In addition, usage of different film preparation processes can also cause significant variations in the material properties and density of designed nanostructure. As a result, current procedures of ultrathin film material properties determination require either simultaneous measurements on multiple sophisticated devices [19–21] or the specially designed micro-/nanomechanical resonators or experimental setup with the advanced computational tools [22–26]. For instance, the high-resolution transmission electron microscope is used to precisely control the force loading/unloading during the nanoindentation of a nanoscale sample [20,21]. Drawbacks of these procedures are that combined measurements on either several sophisticated devices or one specially-designed device are difficult to perform, time consuming, often expensive, and each developed procedure is usually limited to only a specific class of materials. The resonant methods can be also integrated in situ into the nanomaterial deposition systems [27,28].

In response, here we demonstrate the outstanding capability of common microcantilever to determine the density, generated internal stress, the Poisson's ratio, and the elastic properties of solid and polymer ultrathin films, from measured static and dynamic responses of the microcantilever, before and after depositing a thin layer film on its surface. Sketch of the microcantilever with the deposited film is given in Figure 1. We emphasize here that the present procedure utilizes, in addition to well-established measurements of the cantilever static deflection, the flexural and torsional resonant frequencies of the cantilever operating in air; also monitoring often neglected changes in the corresponding quality factors (Q-factors). As a direct consequence, no additional experimental setup or specially designed microcantilevers are required, enabling non-destructive and easily accessible material characterization and testing of ultrathin films. Noticing that Q-factor is a dimensionless parameter describing the efficiency of the designed resonator (i.e., higher Q-factor values stand for lower dissipation and higher efficiency). Importantly, in air the Q-factor is proportional to the material properties and dimensions of the designed microcantilever, and a known air damping that can be precisely controlled by changing the air pressure [29]. Other energy losses, such as the support, surface or the thermo-elastic loss, have only a negligibly small impact on the Q-factor of microcantilever submerged in air. When film is sputtered on the resonator surface, it alters the material properties and dimensions of the microcantilever resonator, yielding changes in Q-factor. As such, Q-factor provides an additional source of information on prepared ultrathin film(s).

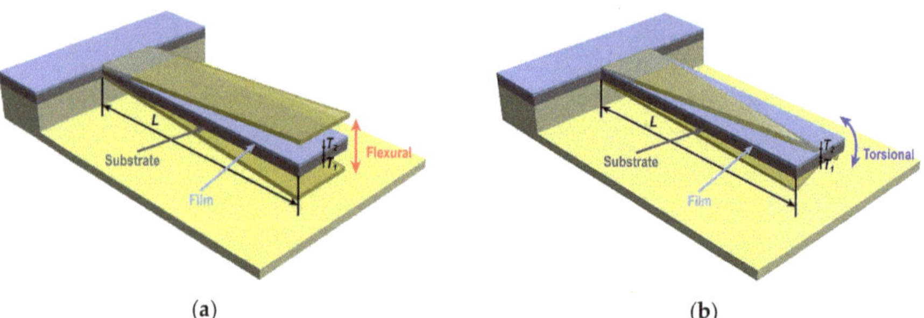

(a)   (b)

Figure 1. Sketch of the two-layered microcantilever made of an elastic substrate and coated ultrathin film performing (**a**) flexural and (**b**) torsional oscillations.

We first derive easily accessible expressions needed to calculate the material properties of an ultrathin film from observed changes in flexural and torsional resonant frequencies and Q-factors of the microcantilever operating in air. Then, we analyze the sensitivity of calculated material properties of ultrathin film on uncertainties in frequency (Q-factor) and dimensions measurements and, afterwards, we validate our theoretical findings by comparing theoretical predictions with experimental results and numerical computations. Despite the fact that analysis is carried out on rectangular microcantilevers, the obtained results and developed procedure of thin film material characterization are valid for other cantilever shapes. In this case, just the flexural and torsional rigidities and hydrodynamic functions used in the model must be recalculated.

2. Theory

2.1. Flexural Oscillations of Two-Layered (Multi-Layered) Microcantilever Operating in Air

To begin, we recall a known fact that once a thin layer film is sputtered on an elastic substrate, it generates in-plane stresses and also alters the overall cantilever resonator elastic properties, particularly in near vicinity of its clamped end [6,30–33]. These effects, that originate from mismatches in strains and the coefficient of thermal expansion between substrate and film, have been proven to notably affect the resonant frequencies of ultrathin cantilever resonators (i.e., thin sheets) [32,33]. Nevertheless, for relatively thick microcantilevers (e.g., ultrathin film sputtered on a thick elastic substrate), of which are considered in the present work, the cantilever free end allows the generated internal stress to be relaxed [31]. Hence, for out-plane flexural vibrational modes, the governing equation for the dynamic deflection function $u(x,t)$ of the microcantilever consisting of substrate and coated film (see Figure 1a) is given by

$$(\rho_1 S_1 + \rho_2 S_2)\frac{\partial^2 u(x,t)}{\partial t^2} + D_F \frac{\partial^4 u(x,t)}{\partial x^4} = F_{\text{drive}}(x,t) + F_{\text{hydro}}(x,t), \tag{1}$$

where $D_F = \frac{1}{12}E_1 W T_1^3 r(\xi_F, \eta)$; subscript 1 and 2 stand for substrate and film, respectively; ρ, S, W, T are the density, cross sectional area, width, and thickness, respectively; $r(\xi_F,\eta) = [\xi_F^2\eta^4 + 4\xi_F\eta(1 + 1.5\eta + \eta^2) + 1]/(1 + \xi_F\eta)$, $\xi_F = E_2/E_1$, $\eta = T_2/T_1$, $F_{\text{drive}}(x,t)$ is the external driving force per unit length of an arbitrary form that set the microcantilever into motion; $F_{\text{hydro}}(x,t)$ is the hydrodynamic force of surrounding air.

It shall be pointed out that detailed theoretical analysis of a homogeneous microcantilever (i.e., made of one material layer) performing flexural oscillations in air can be found in [34,35]. In present work, we extend these theoretical results to account for two material layers required to characterize the material properties of ultrathin films. Moreover, our results can be also directly applied to the microcantilever consisting of N material layers just by recalculating linear density, ρS (where ρ can be viewed as the effective density and S is the cantilever cross-sectional area), and flexural rigidity using the following general relationships:

$$\rho S = \sum_{i=1}^{N} \rho_i S_i, \, D_F = \sum_{i=1}^{N} E_i \int_{\Pi_i} u^{*2} dS - \frac{\left(\sum_{i=1}^{N} E_i \int_{\Pi_i} u^* dS\right)^2}{\sum_{i=1}^{N} E_i S_i}, \tag{2}$$

where u^* is the local coordinate in the lateral direction, Π_i is the i-th region of cantilever beam cross section [36].

The general form of the hydrodynamic force obtained by solving the Fourier-transformed continuity and Navier–Stokes equations (i.e., computations are in the time domain Fourier-transform), for an incompressible fluid as $F_{\text{hydro}}(x|\omega) = \kappa_F \rho_{\text{air}}\omega^2 W^2 \Gamma_F(\omega) U(x|\omega)$, where $U(x|\omega)$ is the Fourier-transformed deflection function, $\kappa_F = \frac{\pi}{4}$, ρ_{air} is the air density, and $\Gamma_F(\omega)$ is the hydrodynamic function for flexural vibration mode. Then, taking the Fourier transform of Equation (1) and rearranging terms yields:

$$\frac{d^4 U(x|\omega)}{dx^4} - \left(\gamma_{(n)}^2 \frac{\omega}{\omega_{vF}^{(n)}}\right)^2 \left[1 + \frac{\kappa_F \rho_{air}\, W}{\rho_1 T_1 (1 + \mu\eta)}\Gamma_F(\omega)\right] U(x|\omega) = \widetilde{F}_{drive}(x|\omega),\qquad(3)$$

where $\mu = \rho_2/\rho_1$, $\widetilde{F}_{drive}(x|\omega) = F_{drive}(x|\omega)L^4/D_F$, $\omega_{vF}^{(n)} = \left(\frac{\gamma_{(n)}}{L}\right)^2 \sqrt{D_F/[\rho_1 T_1(1 + \mu\eta)]}$ is the cantilever angular resonant frequency in vacuum of the n-th vibrational mode, $n = 1, 2, 3, \ldots$ stands for the considered vibrational mode, L is the cantilever length, ω is a characteristic angular frequency of the microcantilever oscillations and $\gamma_{(n)}$ is obtained as the positive root(s) of the following characteristic transcendental equation:

$$\cosh\gamma\,\cos\gamma + 1 = 0.\qquad(4)$$

For an arbitrary form of the driving force the general solution of Equation (3) can be found by the eigenfunction expansion method. In this case, the dynamic deflection function can be obtained as a linear combination of the microcantilever mode shapes, $\theta_{F(n)}(x) = \sinh\left(\gamma_{(n)}x\right) - \sin\left(\gamma_{(n)}x\right) - \left[\frac{\sinh\left(\gamma_{(n)}\right)+\sin\left(\gamma_{(n)}\right)}{\cosh\left(\gamma_{(n)}\right)+\cos\left(\gamma_{(n)}\right)}\right] \times \left[\cosh\left(\gamma_{(n)}x\right) - \cos\left(\gamma_{(n)}x\right)\right]$ (see [37]) and the one reads:

$$U(x|\omega) = \sum_{n=1}^{\infty} B_{F(n)}(\omega)\theta_{F(n)}(x),\qquad(5)$$

where $B_{F(n)}(\omega)$ is found using the orthonormal properties of $\theta_{F(n)}(x)$ as

$$B_{F(n)}(\omega) = \frac{\int_0^L \widetilde{F}_{dr}(\bar{x}|\omega)\theta_{F(n)}(\bar{x})d\bar{x}}{\gamma_{(n)}^4 - \left(\gamma_{(n)}^2 \frac{\omega}{\omega_{vF}^{(n)}}\right)^2 \left[1 + \frac{\kappa_F \rho_{air}\,W}{\rho_1 T_1(1+\mu\eta)}\Gamma_F(\omega)\right]}.\qquad(6)$$

The dissipative effect of air is small compared to viscous fluid (i.e., $Q^{(n)} \gg 1$ [34,35,38]); therefore, in a vicinity of the resonance peaks $\Gamma_F(\omega) \approx \Gamma_{F\_r}(\omega) + i\Gamma_{F\_im}(\omega)$, where $\Gamma_{F\_r}(\omega)$ and $\Gamma_{F\_im}(\omega)$ are the real and imaginary components of the dimensionless hydrodynamic function for cantilever performing flexural oscillations. Then, the resonant frequency and Q-factor of n-th microcantilever vibrational mode in air can be obtained with an analogy to a simple harmonic oscillator as:

$$\omega_F^{(n)} = \frac{\omega_{vF}^{(n)}}{\sqrt{1 + \frac{\kappa_F \rho_{air}\,W}{\rho_1 T_1(1+\mu\eta)}\Gamma_{F\_r}\left(\omega_F^{(n)}\right)}},\qquad(7)$$

$$Q_F^{(n)} = \frac{\frac{\rho_1 T_1(1+\mu\eta)}{\kappa_F \rho_{air}\,W} + \Gamma_{F\_r}\left(\omega_F^{(n)}\right)}{\Gamma_{F\_im}\left(\omega_F^{(n)}\right)}.\qquad(8)$$

For reader's convenience, we present dependencies of the real $\Gamma_{F\_r}\left(\omega_F^{(1)}\right)$ and imaginary $\Gamma_{F\_im}\left(\omega_F^{(1)}\right)$ components of the hydrodynamic function on the fundamental mode frequency represented through the Reynolds number, $Re = \frac{\pi}{4}\omega W^2 \rho_{air}/\mu_{air}$, where μ_{air} is the viscosity of air, in Figure 2a. Noticing only that the Reynolds number is a dimensionless parameter used to predict the flow pattern. For a microcantilever consisting of N material layers, the normalized "effective" density $\rho_1 T_1(1+\mu\eta)$ in Equations (7) and (8) is replaced by $\sum_{i=1}^{N} \rho_i T_i$. Furthermore, for higher vibrational modes the hydrodynamic function depends on the following two dimensionless parameters: (i) The Reynolds number defined now as $Re = \omega W^2 \rho_{air}/\mu_{air}$; and (ii) the normalized mode shape given by $\kappa =$

$\gamma_{(n)}W/L$ [39]. The exact form of the hydrodynamic function for an arbitrary mode number can be found in [39,40]. The general solution for the dynamic deflection has been obtained using the eigenfunction expansion method (see structure of Equation (5)); therefore, the obtained expressions for the resonant frequency and Q-factor represented by Equations (7) and (8) are valid for an arbitrary vibrational mode (i.e., for higher modes only the hydrodynamic function must be recalculated).

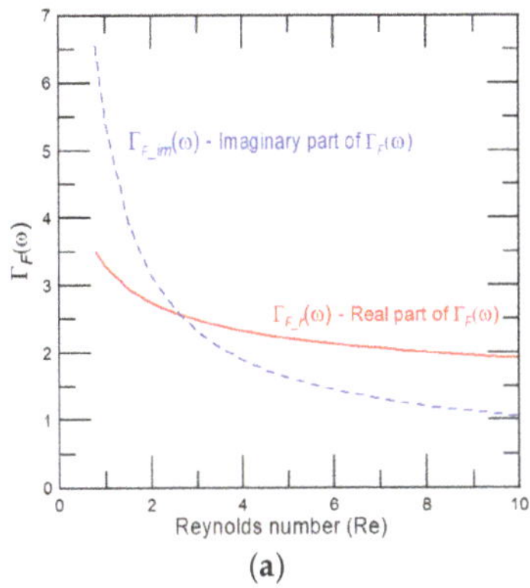

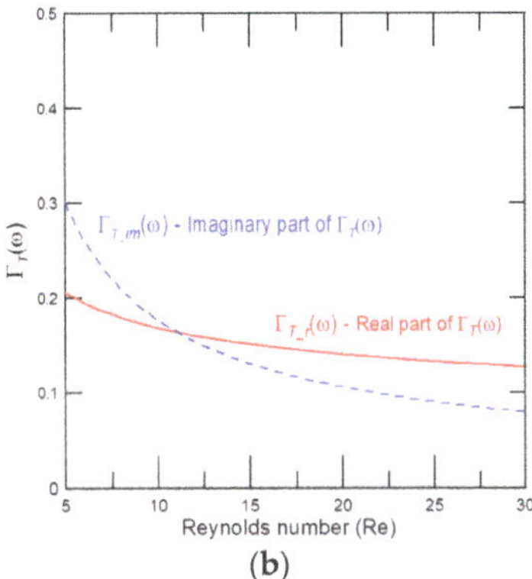

Figure 2. Calculated dependencies of the real and imaginary components of the hydrodynamic function on the fundamental resonant frequency of a rectangular two-layered microcantilever (see Figure 1) immersed in air, represented through the Reynolds number for (**a**) flexural and (**b**) torsional oscillations.

2.2. Torsional Oscillations of Two-Layered (Multi-Layered) Microcantilever Operating in Air

The resonant frequency and Q-factor of a two(multi)-layered microcantilever vibrating in air can be obtained in the same way as done in the previous section for flexural oscillations. Briefly, by accounting for the membrane analogy proposed by Prandtl in 1903, similarities with the theoretical model for flexural oscillations of a multilayered beam in vacuum given in Zapomel et al. [36] and theory of Green and Sader [41], the general governing equation and boundary conditions for torsional oscillations of a two-layered microcantilever (see Figure 1b) operating in air takes the following form:

$$D_{Tr}\frac{\partial^2\phi(x,t)}{\partial x^2} - \frac{\rho_1 W^3 T_1}{12}\left[1 + \varepsilon_1^2 + \mu\eta\left(1 + \varepsilon_1^2\eta^2\right)\right]\frac{\partial^2\phi(x,t)}{\partial t^2} = M_{\text{drive}}(x,t) + M_{\text{hydro}}(x,t), \qquad (9)$$

$$\phi(0,t) = 0, \quad \frac{\partial\phi(L,t)}{\partial x} = 0. \qquad (10)$$

Here $\phi(x, t)$ is the deflection angle about the cantilever major axis, $\varepsilon_1 = T_1/W$ is the characteristic dimensional scale, $D_{Tr} = \frac{1}{3}G_1 WT_1^3 r(\xi_{Tr}, \eta)$ is the torsional rigidity, $r(\xi_{Tr}, \eta) = [\xi_{Tr}^2\eta^4 + 4\xi_{Tr}\eta(1 + 1.5\eta + \eta^2) + 1]/(1 + \xi_{Tr}\eta)$, $\xi_{Tr} = G_2/G_1$, G is the shear modulus; $M_{\text{drive}}(x,t)$ is the driving moment per unit length and $M_{\text{hydro}}(x,t)$ is the hydrodynamic torque per unit length, which the general form is obtained by solving the equation of motion of fluid in complex space [39,41], given by:

$$M_{\text{hydro}}(x|\omega) = -\frac{\pi}{8}\rho_{\text{air}}\omega^2 W^4\Gamma_T(\omega)\Phi(x|\omega), \qquad (11)$$

where $\Gamma_T(\omega)$ is the torsional dimensionless hydrodynamic function, $\Phi(x|\omega)$ is the deflection angle in complex space. Plugging Equation (11) into the Fourier-transformed Equation (9) and rearranging terms yields:

$$\frac{d^2\Phi(x|\omega)}{dx^2} - \left(\lambda_{(n)}\frac{\omega}{\omega_{vTr}^{(n)}}\right)^2\left[1 + \frac{\kappa_T\rho_{air}W}{\rho_1 T_1\left[1 + \varepsilon_1^2 + \mu\eta\left(1 + \varepsilon_1^2\eta^2\right)\right]}\Gamma_T(\omega)\right]\Phi(x|\omega) = \widetilde{M}_{drive}(x|\omega), \qquad (12)$$

where $\kappa_T = 3\pi/2$, $\widetilde{M}_{drive}(x|\omega) = M_{drive}(x|\omega)/D_{Tr}$, $\omega_{vTr}^{(n)} = \frac{\lambda_{(n)}}{L}\sqrt{\frac{4G_1 T_1^2 r(\xi_{Tr},\eta)}{\rho_1 W^2\left[1 + \varepsilon_1^2 + \mu\eta\left(1 + \varepsilon_1^2\eta^2\right)\right]}}$ is the angular resonant frequency in vacuum of the n-th torsional mode and $\lambda_{(n)} = \pi(2n-1)/2$, $n = 1, 2, 3, \ldots$.

The general solution for torsional oscillations of the microcantilever driven by an arbitrary form torque can be again obtained by the eigenfunction expansion method:

$$\Phi(x|\omega) = \sum_{n=1}^{\infty} B_{Tr(n)}(\omega)\theta_{Tr(n)}(x), \qquad (13)$$

where $\theta_{Tr(n)}(x) = \sin[(2n-1)\pi x/2]$, $n = 1, 2, 3, \ldots$ and $B_{Tr(n)}(\omega)$ is given by

$$B_{Tr(n)}(\omega) = \frac{2\int_0^L \widetilde{M}_{dr}(\overline{x}|\omega)\theta_{Tr(n)}(\overline{x})d\overline{x}}{\left(\lambda_{(n)}\frac{\omega}{\omega_{vTr}^{(n)}}\right)^2\left[1 + \frac{\kappa_T\rho_{air}W}{\rho_1 T_1\left[1 + \varepsilon_1^2 + \mu\eta\left(1 + \varepsilon_1^2\eta^2\right)\right]}\Gamma_T(\omega)\right] - \lambda_{(n)}^2}. \qquad (14)$$

Then, in analogy with flexural motion, for small dissipative effects the desired expressions that enable to accurately predict the torsional resonant frequency and Q-factor of the n-th vibrational mode of microcantilever consisting of substrate and ultrathin film (i.e., $\eta \ll 1$ and $\varepsilon_1 < 1$) operating in air are:

$$\omega_{Tr}^{(n)} \approx \frac{\omega_{vTr}^{(n)}}{\sqrt{1 + \frac{\kappa_T\rho_{air}W}{\rho_1 T_1(1+\mu\eta)}\Gamma_{T\_r}\left(\omega_{Tr}^{(n)}\right)}} \qquad (15)$$

and

$$Q_{Tr}^{(n)} \approx \frac{\frac{\rho_1 T_1(1+\mu\eta)}{\kappa_T\rho_{air}W} + \Gamma_{T\_r}\left(\omega_{Tr}^{(n)}\right)}{\Gamma_{T\_im}\left(\omega_{Tr}^{(n)}\right)}, \qquad (16)$$

where $\Gamma_{T\_r}\left(\omega_{Tr}^{(n)}\right)$ and $\Gamma_{T\_im}\left(\omega_{Tr}^{(n)}\right)$ are the real and imaginary components of the dimensionless hydrodynamic function that, for fundamental torsional vibrational mode, are given in Figure 2b.

For a multilayered beam, just coefficients for torsional mass and hydrodynamic effect of fluid must be recalculated. In general cases, $\rho_1 T_1(1 + \mu\eta)$ is replaced by $\sum_{i=1}^{N}\rho_i I_{Pi}$, where I_p is the polar moment of inertia and the hydrodynamic effect of air is now represented by $\left(\frac{\pi}{8}\right)\rho_{air}W^4$ [see Equation (11)].

3. Results

3.1. Method of Determining Material Properties, Density, and Generated Internal Stress of Ultrathin Film(s)

We first evaluate impact of coated film on the microcantilever resonant frequency and Q-factor and, afterwards, we derive easily accessible expressions enabling calculation of the density, the Young's and shear moduli, the Poisson's ratio and stress of the solid and polymer ultrathin films from experimentally observed changes in the microcantilever resonant frequencies, Q-factors, and the cantilever static deflection. Using Equations (7), (8), (15), and (16), the desired changes in resonant frequency and

Q-factor of the n-th vibrational mode, represented by ratio of the microcantilever made of substrate and coated film (i.e., ω and Q) to the one of without film (i.e., ω_0 and Q_0), are obtained as

$$\frac{\omega_j^{(n)}}{\omega_{j0}^{(n)}} = \sqrt{r(\xi_j,\eta) \times \left[\frac{1 + C_j \Gamma_{j\_r}\left(\omega_{j0}^{(n)}\right)}{1 + \mu\eta + C_j \Gamma_{j\_r}\left(\omega_j^{(n)}\right)}\right]}, \tag{17}$$

$$\frac{Q_j^{(n)}}{Q_{j0}^{(n)}} = \frac{\Gamma_{j\_im}\left(\omega_{j0}^{(n)}\right)}{\Gamma_{j\_im}\left(\omega_j^{(n)}\right)} \times \left[\frac{1 + \mu\eta + C_j \Gamma_{j\_r}\left(\omega_j^{(n)}\right)}{1 + C_j \Gamma_{j\_r}\left(\omega_{j0}^{(n)}\right)}\right], \tag{18}$$

where subscript $j = F$ (flexural) and Tr (torsional), and $C_j = \kappa_j \frac{\rho_{air}}{\rho_1} \frac{W}{T_1}$.

Once film is coated on the elastic substrate, it alters the microcantilever flexural and torsional rigidity represented by the dimensionless parameter $r(\xi_j,\eta)$ and increases its "effective" linear density through the coefficient $\mu\eta$. According to Equations (17) and (18), changes in the microcantilever resonant frequency caused by the film differ essentially from those obtained for Q-factor of the same vibrational mode. Since dissipative effect of air is small [38], changes in the resonant frequency depend just on interplay between the rigidity and effective linear density of the prepared sample as $r(\xi_j,\eta)/(1 + \mu\eta)$. As a result, an increase, decrease, or even non-monotonic dependency of ω/ω_0 on film thickness can be observed depending on the exact film and substrate material properties and density values. Quality factor is; however, proportional to combination of the linear density and known hydrodynamic load represented by Γ_{j_r} and Γ_{j_im}. We note that rigidity affects the Q-factor only indirectly through the resonant frequency used to calculate both components of the hydrodynamic function (see Equation (18)). As such, with an increase of film thickness only an increase in Q-factor can be observed. It immediately implies that combined measurements of the resonant frequency and Q-factor changes enable evaluation of the material properties of ultrathin film, even when no shift in the resonant frequency can be observed [24]. For example, dependencies of the fundamental resonant frequency and Q-factor changes of the silicon microcantilever of length $L = 300$ μm, $W = 30$ μm, and $T_1 = 1$ μm ($\rho_1 = 2.33$ g/cm^3, $E_1 = 169$ GPa, and $G_1 = 42$ GPa) on thickness of film made of gold ($\rho_2 = 19.3$ g/cm^3, $E_2 = 79$ GPa, and $G_2 = 27$ GPa), platinum ($\rho_2 = 21.45$ g/cm^3, $E_2 = 168$ GPa, and $G_2 = 61$ GPa), and silicon nitride ($\rho_2 = 3.2$ g/cm^3, $E_2 = 350$ GPa, and $G_2 = 100$ GPa) are given in Figure 3. As expected, an increase and/or decrease in the frequency ratio can be observed depending on the density and material properties of coated film (Figure 3a,b). For gold and platinum (silicon nitride), film density (rigidity) dominates the frequency response, thus with an increase of film thickness the resonance shifts to lower (higher) values. For a given film thickness, the higher Q/Q_0 values can be achieved for heavier films (i.e., platinum and gold; Figure 3c,d). Importantly, obtained theoretical predictions are in a good agreement with published experimental observations carried out on the microcantilever resonator-based biosensor [42]. In these experiments, the antibody and antigen formed thin layer films on the cantilever surface, yielding both an increase and decrease in the resonant frequency depending on the interplay between stiffness and stress effects, and just an increase in Q-factor as predicted by the present model.

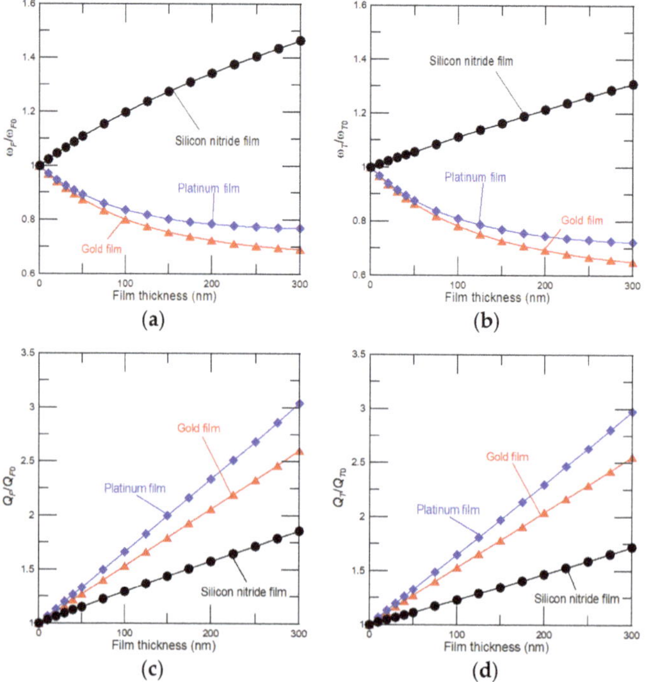

Figure 3. Dependencies of (**a**,**b**) the fundamental resonant frequencies and, correspondingly, (**c**,**d**) Q-factors of the silicon cantilever on film thickness for flexural and torsional vibrational modes.

Rearranging terms in Equations (17) and (18), the density, and the Young's and shear moduli of prepared ultrathin film can be determined from the following equations:

$$\mu = \frac{1}{\eta} \frac{Q_j^{(n)}}{Q_{j0}^{(n)}} \frac{\Gamma_{j\_im}\left(\omega_j^{(n)}\right)}{\Gamma_{j\_im}\left(\omega_{j0}^{(n)}\right)} \left(1 + C_j \Gamma_{j\_r}\left(\omega_{i0}^{(n)}\right)\right) - \frac{1}{\eta}\left(1 + C_j \Gamma_{j\_r}\left(\omega_j^{(n)}\right)\right), \tag{19}$$

$$r\left(\xi_j, \eta\right) = \left(\frac{\omega_j^{(n)}}{\omega_{j0}^{(n)}}\right)^2 \frac{Q_j^{(n)}}{Q_{j0}^{(n)}} \frac{\Gamma_{j\_im}\left(\omega_j^{(n)}\right)}{\Gamma_{j\_im}\left(\omega_{j0}^{(n)}\right)}. \tag{20}$$

The Young's modulus is related to the shear modulus as $E = 2G(1 + v)$, where v is the Poisson's ratio. Hence, accounting for Equation (20), the Poisson's ratio of coated film is then calculated by:

$$v_2 \approx \frac{(R_F - 1)B_T}{(R_T - 1)B_F}(1 + v_1) - 1, \tag{21}$$

where $R_j = \left(\frac{\omega_j^{(n)}}{\omega_{j0}^{(n)}}\right)^2 \frac{Q_j^{(n)}}{Q_{j0}^{(n)}} \frac{\Gamma_{j\_im}\left(\omega_j^{(n)}\right)}{\Gamma_{j\_im}\left(\omega_{j0}^{(n)}\right)}$ and $B_j = 4 + 6\eta + 4\eta^2 - R_j$. And, finally, the generated internal stress can be obtained from the microcantilever static deflection measurement [7]:

$$\sigma = \frac{\left[(1 - v_2) + (1 - v_1)\xi_F \eta^3\right]E_1 T_1}{3(1 - v_2)(1 - v_1)(1 + \eta)\eta L^2} z, \tag{22}$$

where z is the detected microcantilever static deflection caused by stress due to film coating. Equation (22) is obtained using a plate approximation and without accounting for the clamped end effect. Hence, this equation is strictly valid for microcantilevers with a large aspect ratio $L/W \gg 5$. If the aspect ratio is small, then the effect of clamping region must be accounted and the microcantilever deflection can be obtained by following approach given in work of Tamayo et al. [43]. In present work we assume $L \gg W \gg T$ and, consequently, Equation (22) describes accurately the relationship between bending of the cantilever free end and the internal stress [44].

Equations (19) and (20) reveal that the Young's and shear moduli of prepared film can be determined even without a requirement for knowing its density and vice versa. Similarly, the Poisson's ratio of ultrathin film calculated by Equation (21) does not require previous calculation of the Young's and shear moduli of film and substrate. These findings are particularly of value in testing of micro-/nano-electronic devices, where in order to prevent their mechanical failure, it is of emergent importance to know the material properties of designed ultrathin films [45]. It shall be pointed out that the present method can be also extended to determine, in addition to elastic properties and density, the ultrathin film thickness. In this case, the rarely-measured in-plane flexural resonant frequencies must be taken into account [24]. Then, for in-plane flexural mode $r(\xi_F, \eta) = 1 + \xi_F \eta$ and, consequently, the density, elastic properties and thickness of prepared film can be determined using Equations (19) and (20).

3.2. Impact of Errors in Dimensions, Frequency, and Quality Factor Measurements on the Accuracy of the Present Method

To ensure the proposed procedure of material properties determination is practical, we now examine impact of the dimensional discrepancy and uncertainties in frequency measurements on the determined material properties. It is worth noting that thickness of film can be measured by the ellipsometry with a typical measurement error of sub-nanometer, whereas the cantilever length and width are often determined by a scanning electron microscope with the common uncertainties ranging from few nm to tens of nm. In general, the uncertainties in dimensions, frequency, and Q-factor yield inaccuracies in the determined properties of designed ultrathin films. These inaccuracies expressed through errors in the dimensionless thickness, $\Delta\eta$, density, $\Delta\mu$, modulus parameters, $\Delta\xi_i$, and the Poisson's ratio, $\Delta\upsilon$, can be viewed as a perturbed term in a given quantity. The relative errors calculated from Equations (19)–(21) for $\eta \ll 1$ (i.e., $r(\xi_j, \eta) \approx [4\xi_j\eta(1+ 1.5\eta) + 1]/(1 + \xi_j\eta))$ read:

$$\frac{\Delta\mu}{\mu} = \frac{1}{1 + \frac{\Delta\eta}{\eta}} - 1, \tag{23}$$

$$\frac{\Delta\xi_i}{\xi_i} = \frac{1}{1 + \frac{6\Delta\eta}{6\eta+4-R_i^*} + \frac{\Delta\eta}{\eta}} - 1, \tag{24}$$

$$\frac{\Delta\overline{\upsilon}}{\overline{\upsilon}} = \frac{1 + 4\Delta\eta(1.5 + 2\eta)/B_T^*}{1 + 4\Delta\eta(1.5 + 2\eta)/B_F^*} - 1, \tag{25}$$

where $\overline{\upsilon} = (1 + \upsilon_2)/(1 + \upsilon_1)$, R_i^* and B_i^* are the measured and calculated properties with due account for uncertainties in the frequency, Q-factor, and dimensions. The achievable relative sensitivity in determined material properties and density of gold film sputtered on the silicon substrate, with dimensions 300 µm (L), 30 µm (W), and 1 µm (T_1), are given in Figure 4. For example, for 40 nm thick gold film and the uncertainties in frequency, thickness, and width measurements of 0.5 kHz, 1 and 10 nm, the following properties of gold film are obtained: The Young's modulus of 78.9 ± 3.3 GPa, the shear modulus of 27.2 ± 0.4 GPa, the density of 20.3 ± 1.2 g/cm^3, and the Poisson's ratio of 0.45 ± 0.1. These results demonstrate that the present procedure of material properties measurement is accurate, even for the relatively high uncertainties in thickness and resonant frequencies (Q-factors) measurements. Importantly, for given measurement errors, the accuracy in determined material properties can be easily improved just by detecting changes in the resonant frequency and Q-factor of the higher vibrational modes (i.e., higher resonant frequencies yield a significant increase in Q-factor values) [38]. For an illustration, we again extract the material properties of 40 nm thick gold film with errors in measurements given in the previous example by considering the second vibrational mode. The calculated properties of gold film using the second vibrational modes are as follows: The Young's modulus of 78.8 ± 1.4 GPa, the shear modulus of 27.2 ± 0.2 GPa, the density of 20.1 ± 0.9 g/cm^3, and the Poisson's ratio of 0.43 ± 0.1. Different values of film's Young's modulus, density and the Poisson's ratio mainly originate from the uncertainties in calculated $\Gamma_{j\_r}\left(\omega_j^{(n)}\right)$ and $\Gamma_{j\_im}\left(\omega_j^{(n)}\right)$ [34,39].

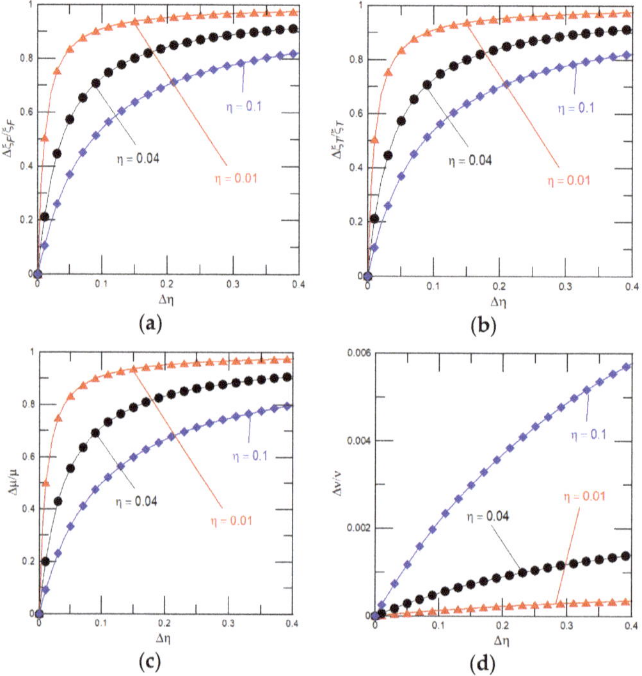

Figure 4. The achievable relative sensitivity of (**a**) the Young's modulus (Equation (24)); (**b**) the shear modulus (Equation (24)); (**c**) density (Equation (23)); and (**d**) the Poisson's ratio (Equation (25)) of gold film of $\eta = 0.01$, 0.04, and 0.1 coated on the silicon substrate with dimensions 300 μm (L), 30 μm (W), and 1 μm (T_1).

4. Discussion

We now assess the validity and versatility of the proposed procedure of material characterization by extracting the Young's modulus of the atomic-layer-deposited TiO$_2$ ultrathin film, of thickness 20 and 50 nm, sputtered on the microcantilever made of SU-8 photoresist polymer substrate [46]. In contrast to data presented in [46], where the density of TiO$_2$ film was estimated based on the X-ray reflectometry measurements and, then, the Young's modulus of TiO$_2$ film was calculated from changes in the resonant frequencies before and after conformal coating of the film, we determine the TiO$_2$ film Young's modulus without requirement for the density measurement. We remind the reader that the effective linear density of microcantilever vibrating in air can be expressed through measured Q-factor values and, consequently, the simple flexural resonant frequencies of the n-th vibrational mode can be accurately predicted by:

$$f_{F0}^{(n)} = \frac{\gamma_{(n)}^2}{2\pi L^2}\sqrt{\frac{1}{\kappa_F \rho_{\text{air}}}\frac{D_F}{W^2}\frac{D_F}{Q_F^{(n)}\Gamma_{F\_im}(\omega)}}, \qquad (26)$$

where $f = \omega/(2\pi)$ and the flexural rigidity of a microcantilever with conformal coating (i.e., the TiO_2 film covers entire surface area of the microcantilever) is given by:

$$D_F = \frac{1}{12}E_1 WT_1^3 + \frac{1}{6}E_2\left[T_1^3 T_2 + T_2^3(W + 2T_2) + 3T_2(W + 2T_2)(T_1 + T_2)^2\right]. \tag{27}$$

For a microcantilever of length, width, and substrate thickness of 300 ± 1 μm, 100 ± 1 μm, and 5.6 ± 0.05 μm, substrate density of 1.2 ± 0.01 g/cm^3, and the measured fundamental resonant frequency in air of $f_{F0}^{(1)} = 18.2 \pm 0.32$ kHz, the Young's modulus of SU-8 substrate of 4.04 ± 0.3 GPa is estimated by Equation (26). Importantly, published experimental data from [46] shows that, in accordance with present theoretical predictions (see Figure 3a,c), coated TiO_2 film shifts the microcantilever resonant frequencies to the higher values (i.e., for TiO_2 film, stiffness dominates and causes an increase in Q-factor). Accounting for the flexural rigidity given by Equation (27) and structure of Equation (20), the expression enabling calculation of the Young's modulus of TiO_2 film from the experimentally-detected changes in frequency response reads:

$$\xi_F = \frac{1}{2r_0}\left[\left(\frac{f_F^{(n)}}{f_{F0}^{(n)}}\right)^2 \frac{Q_F^{(n)}}{Q_{F0}^{(n)}} \frac{\Gamma_{F\_im}\left(\omega_F^{(n)}\right)}{\Gamma_{F\_im}\left(\omega_{F0}^{(n)}\right)} - 1\right], \tag{28}$$

where $r_0 = \varepsilon_2 + \eta^3(1 + 2\varepsilon_2) + 3\eta(1 + 2\varepsilon_2)(1 + \eta)^2$ and $\varepsilon_2 = T_2/W$. Table 1 presents a comparison of the Young's modulus of 20 and 50 nm thick TiO_2 films calculated using Equation (28) and determined previously by Colombi et al. [46].

Table 1. Comparison of the Young's modulus of atomic layer deposition (ALD) TiO_2 films calculated by Equation (28) and determined previously in [46].

Measured Quantity	SU-8	20 nm TiO$_2$	50 nm TiO$_2$
Frequency in air (kHz)	18.2 ± 0.32	21.0 ± 0.21	27.1 ± 0.18
Q_F/Q_{F0}	None	1.11 ± 0.018	1.34 ± 0.019
Young's modulus (GPa), by Equation (28)	4.04 ± 0.30	66.3 ± 14	96.9 ± 8
Young's modulus (GPa), [34]	3.82 ± 0.13	60 ± 18	91 ± 15

In addition, comparisons of the shear moduli and densities of 20 and 50 nm thick silicon nitride (Si_3N_4) films, coated on the silicon microcantilevers of length 300 μm, width 30 μm, and thickness of 1 μm, calculated by the present method (i.e., Equations (19) and (20)) and numerically by using COMSOL Multiphysics, are given in Table 2. The uncertainties in frequency, thickness, and width measurements are: 0.5 kHz, 1, and 10 nm, respectively.

Table 2. Comparisons of the shear modulus and density of silicon nitride film coated on the silicon microcantilever obtained by the proposed procedure and numerically. Considered errors in frequency, thickness, and width measurements are: 0.5 kHz, 1, and 10 nm, respectively.

Measured Quantity	20 nm Si$_3$N$_4$	50 nm Si$_3$N$_4$
ω_T/ω_{T0}, Equation (17)/Numerically	$1.024 \pm 0.003/1.026 \pm 0.003$	$1.058 \pm 0.003/1.061 \pm 0.003$
Q_T/Q_{T0}, Equation (18)/Numerically	$1.045 \pm 0.002/1.048 \pm 0.002$	$1.114 \pm 0.002/1.116 \pm 0.002$
Density (g/cm^3), Equation (19)/Numerically	$3.35 \pm 0.01/3.52 \pm 0.02$	$3.25 \pm 0.01/3.26 \pm 0.01$
Shear modulus (GPa), Equation (20)/Numerically	$100.48 \pm 10.77/107.64 \pm 11.11$	$100.39 \pm 4.67/104.01 \pm 4.74$

The results given in Tables 1 and 2 show that here derived expressions enabling calculation of the ultrathin film material properties and density are valid and, in addition, the proposed procedure of material properties determination is relatively simple, practical, universal, and accurate, even for low accuracies in dimensions and frequency measurements.

5. Conclusions

We have proposed and demonstrated the non-destructive and easily accessible method of material characterization utilizing the well-established measurement of static and dynamic modes of the microcantilevers operating in air. Expressions needed to calculate the film properties from measured frequency and Q-factor changes are derived. We have showed that by monitoring changes in resonant frequency and, correspondingly, Q-factor, the Young's (shear) modulus of film can be determined without the requirement of knowing the film density. This finding would be of great value in material testing of ultrathin films of which the density deviates from known bulk values. The usual discrepancies in dimensions and errors in frequency (Q-factor) measurements were proven to have only a small impact on the calculated material properties of ultrathin film. In addition, for given errors in dimensions and frequency measurements, the accuracy in extracted film properties can be easily improved by using the higher vibrational modes. A good agreement between the Young's modulus (the shear modulus and density) determined by the present procedure and previous experimental measurements (numerical computations) carried out on the microcantilever, consisting of an elastic substrate and coated ultrathin film(s), has allowed us to confirm the validity of: (a) derived expressions and (b) the present procedure of the ultrathin film material properties determination.

Author Contributions: I.S. proposed the topic, developed the models, and wrote the manuscript. I.S. and L.G. performed the systematic investigations and analysis of data.

Funding: This work was supported by Harbin Institute of Technology, Shenzhen, Shenzhen, China, the project of Czech Republic SOLID 21 (CZ.02.1.01/0.0/0.0/16_019/0000760) and Drážní revize s.r.o.

Conflicts of Interest: The authors declare no conflicts of interest.

References

1. Cerdán-Pasarán, A.; López-Luke, T.; Esparza, D.; Zarazúa, I.; De la Rosa, E.; Fuentes-Ramírez, R.; Alatorre-Ordaz, A.; Sánchez-Solís, A.; Torres-Castro, A.; Zhang, J.Z. Photovoltaic properties of multilayered quantum dot/quantum rod-sensitized TiO_2 solar cells fabricated by SILAR and electrophoresis. *Phys. Chem. Chem. Phys.* **2015**, *17*, 18590–18599. [CrossRef] [PubMed]
2. Yunin, P.A.; Drozdov, Y.N.; Drozdov, M.N.; Korolev, S.A.; Lobanov, D.N. Study of multilayered SiGe semiconductor structures by X-ray diffractometry, grazing-incidence X-ray reflectometry, and secondary-ion mass spectrometry. *Semiconductors* **2013**, *47*, 1556–1561. [CrossRef]
3. Matsuda, T.; Ishihara, H. Proposal of highly efficient photoemitter with strong phonon-harvesting capability and exciton supperradiance. *Appl. Phys. Lett.* **2017**, *111*, 063108. [CrossRef]
4. Li, D.; Jiang, Y.; Zhang, P.; Shan, D.; Xu, J.; Li, W.; Chen, K. The phosphorus and boron co-doping behaviors at nanoscale in Si nanocrystals/SiO_2 multilayers. *Appl. Phys. Lett.* **2017**, *110*, 233105. [CrossRef]
5. Stachiv, I.; Sittner, P.; Olejnicek, J.; Landa, M.; Heller, L. Exploiting NiTi shape memory alloy films in design of tunable high frequency microcantilever resonators. *Appl. Phys. Lett.* **2017**, *111*, 213105. [CrossRef]
6. Stachiv, I.; Sittner, P. Nanocantilevers with adjustable static deflection and significantly tunable spectrum resonant frequencies for applications in nanomechanical mass sensors. *Nanomaterials* **2018**, *8*, 106. [CrossRef] [PubMed]
7. Stachiv, I.; Fang, T.-H.; Jeng, Y.-R. Mass detection in viscous fluid utilizing vibrating micro- and nanomechanical mass sensors under applied axial tensile force. *Sensors* **2015**, *15*, 19351–19368. [CrossRef]
8. Ceccacci, A.C.; Chen, C.; Hwu, E.; Morelli, L.; Bose, S.; Bosco, F.G.; Schmid, S.; Boisen, A. Blu-Ray-based micromechanical characterization platform for biopolymer degradation assessment. *Sens. Actuators B Chem.* **2017**, *241*, 1303–1309. [CrossRef]
9. Stachiv, I.; Fedorchenko, A.I.; Chen, Y.L. Mass detection by means of the vibrating nanomechanical resonators. *Appl. Phys. Lett.* **2012**, *100*, 093110. [CrossRef]
10. Hanay, M.S.; Kelber, S.; Naik, A.K.; Chi, D.; Hentz, S.; Bullard, E.C.; Colinet, E.; Duraffourg, L.; Roukes, M.L. Single-protein nanomechanical mass spectrometry in real time. *Nat. Nanotechnol.* **2012**, *7*, 602–608. [CrossRef]

11. Sage, E.; Sansa, M.; Fostner, S.; Defoort, M.; Gély, M.; Naik, A.K.; Morel, R.; Duraffourg, L.; Roukes, M.L.; Alava, T.; et al. Single-particle mass spectrometry with arrays of frequency-addressed nanomechanical resonators. *Nat. Commun.* **2018**, *9*. [CrossRef] [PubMed]

12. Stachiv, I. Impact of surface and residual stresses and electro-/magnetostatic axial loading on the suspended nanomechanical based mass sensors: A theoretical study. *J. Appl. Phys.* **2014**, *115*, 214310. [CrossRef]

13. Kim, M.G.; Kanatzidis, M.G.; Faccheti, A.; Marks, T.J. Low temperature fabrication of high-performance metal oxide thin-film electronics via combustion process. *Nat. Mater.* **2011**, *10*, 382–388. [CrossRef] [PubMed]

14. Oliver, W.C.; Pharr, G.M. An improved technique for determining hardness and elastic modulus using load and displacement sensing indentation experiments. *J. Mater. Res.* **1992**, *7*, 1564–1583. [CrossRef]

15. Vlassak, J.J.; Nix, W.D. A new bulge test technique for the determination of Young's modulus and Poisson's ratio of thin film. *J. Mater. Res.* **1992**, *7*, 3242–3249. [CrossRef]

16. Srikar, V.T.; Spearing, S.M. A critical review of microscale mechanical testing methods used in the design of micromechanical systems. *Exp. Mech.* **2003**, *43*, 238–247. [CrossRef]

17. Xiang, H.F.; Xu, Z.X.; Roy, V.A.L.; Che, C.M.; Lai, T.P. Method for measurement of the density of thin films of small organic molecules. *Rev. Sci. Instrum.* **2007**, *78*, 034104. [CrossRef] [PubMed]

18. Xu, J.; Umehara, H.; Kojima, I. Effect of deposition parameters on composition, structures, density and topography of CrN films deposited by r.f. magnetron sputtering. *Appl. Surf. Sci.* **2002**, *201*, 208–218. [CrossRef]

19. Whiteside, P.J.D.; Chinisis, J.A.; Hunt, H.K. Techniques and challenges for characterizing metal thin films with application in photonics. *Coatings* **2016**, *6*, 35. [CrossRef]

20. Tsai, P.C.; Jeng, Y.R.; Lee, J.T.; Stachiv, I.; Sittner, P. Effects of carbon nanotube reinforcement and grain size refinement mechanical properties and wear behaviors of carbon nanotube/copper composites. *Diam. Relat. Mater.* **2017**, *74*, 197–204. [CrossRef]

21. Carlton, C.E.; Ferreira, P.J. In Situ TEM nanoindentation of nanoparticles. *Micron* **2012**, *43*, 1134–1139. [CrossRef] [PubMed]

22. Ma, S.; Huang, H.; Lu, M.; Veidt, M. A simple resonant method that can simultaneously measure elastic modulus and density of thin films. *Surf. Coat. Technol.* **2012**, *209*, 208–211. [CrossRef]

23. Stachiv, I.; Zapomel, J.; Chen, Y.L. Simultaneous determination of the elastic modulus and density/thickness of ultrathin films utilizing micro-/nanoresonators under applied axial force. *J. Appl. Phys.* **2014**, *115*, 124304. [CrossRef]

24. Illic, B.; Krylov, S.; Craighead, H.C. Young's modulus and density measurements of thin atomic layer deposited films using resonant nanomechanics. *J. Appl. Phys.* **2010**, *108*, 044317. [CrossRef]

25. Stachiv, I.; Vokoun, D.; Jeng, Y.R. Measurement of Young's modulus and volumetric mass density / thickness of ultrathin films utilizing resonant based mass sensors. *Appl. Phys. Lett.* **2014**, *104*, 083102. [CrossRef]

26. Dufrene, Y.F.; Martinez-Martin, D.; Medalsy, I.; Alsteens, D.; Muller, D.J. Multiparametric imaging of biological systems by force-distance curve–based AFM. *Nat. Methods* **2013**, *10*, 847–854. [CrossRef] [PubMed]

27. Moreland, J. Micromechanical instruments for ferromagnetic measurements. *J. Phys. D Appl. Phys.* **2003**, *36*, R39–R51. [CrossRef]

28. Hwang, K.S.; Lee, J.H.; Park, J.; Yoon, D.S.; Park, J.H.; Kim, T.S. In-Situ quantitative analysis of a prostate-specific antigen (PSA) using a nanomechanical PZT cantilever. *Lab Chip* **2004**, *4*, 547–552. [CrossRef]

29. Sandberg, R.; Mølhave, K.; Boisen, A.; Svendsen, W. Effect of gold coating on the Q-factor of a resonant cantilever. *J. Micromech. Microeng.* **2005**, *15*, 2249–2253. [CrossRef]

30. Lachut, M.J.; Sader, J.E. Effect of surface stress on the stiffness of cantilever plates. *Phys. Rev. Lett.* **2007**, *99*, 206102. [CrossRef]

31. Karabalin, R.B.; Villanueva, L.G.; Matheny, M.H.; Sader, J.E.; Roukes, M.L. Stress-induced variation in the stiffness of micro- and nanocantilever beams. *Phys. Rev. Lett.* **2012**, *108*, 236101. [CrossRef] [PubMed]

32. Pini, V.; Ruz, J.J.; Kosaka, P.M.; Malvar, O.; Calleja, M.; Tamayo, J. How two-dimensional bending can extraordinary stiffen thin sheets. *Sci. Rep.* **2016**, *6*, 29627. [CrossRef] [PubMed]

33. Ruz, J.J.; Pini, V.; Malvar, O.; Kosaka, P.M.; Calleja, M.; Tamayo, J. Effect of surface stress induced curvature on the eigenfrequencies of microcantilever plates. *AIP Adv.* **2018**, *8*, 105213. [CrossRef]

34. Sader, J.E. Frequency response of cantilever beams immersed in viscous fluids with applications to the atomic force microscope. *J. Appl. Phys.* **1998**, *84*, 64–76. [CrossRef]

35. Stachiv, I.; Fang, T.H.; Chen, T.H. Micro-/nanosized cantilever beams and mass sensors under applied axial tensile/compressive force vibrating in vacuum and viscous fluid. *AIP Adv.* **2015**, *5*. [CrossRef]

36. Zapomel, J.; Stachiv, I.; Ferfecki, P. A novel method combining Monte Carlo–FEM simulations and experiments for simultaneous evaluation of the ultrathin film mass density and Young's modulus. *Mech. Syst. Signal Process.* **2016**, *66–67*, 223–231. [CrossRef]

37. Kelly, S.G. *Fundamentals of Mechanical Vibrations*, 2nd ed.; McGraw-Hill International: Singapore, 2000.

38. Chon, J.W.M.; Mulvaney, P.; Sader, J.E. Experimental validation of theoretical models for the frequency response of atomic force microscope cantilever beams immersed in fluids. *J. Appl. Phys.* **2000**, *87*, 3978–3988. [CrossRef]

39. Van Eysden, C.A.; Sader, J.E. Frequency response of cantilever beams immersed in viscous fluids with applications to the atomic force microscope: Arbitrary mode number. *J. Appl. Phys.* **2007**, *101*, 044908. [CrossRef]

40. Van Eysden, C.A.; Sader, J.E. Small amplitude oscillations of a flexible thin blade in a viscous fluid: Exact analytical solution. *Phys. Fluids* **2006**, *18*, 123102. [CrossRef]

41. Green, C.P.; Sader, J.E. Torsional frequency response of cantilever beams immersed in viscous fluids with applications to the atomic force microscope. *J. Appl. Phys.* **2002**, *92*, 6262–6274. [CrossRef]

42. Gupta, A.K.; Nair, P.R.; Akin, D.; Ladisch, M.R.; Broyles, S.; Alam, M.A.; Bashir, R. Anomalous resonance in a nanomechanical biosensor. *Proc. Natl. Acad. Sci. USA* **2006**, *103*, 13362–13367. [CrossRef] [PubMed]

43. Tamayo, J.; Ruz, J.J.; Pini, V.; Kosaka, P.; Calleja, M. Quantification of the surface stress in microcantilever biosensors: Revisiting Stoney's equation. *Nanotechnology* **2012**, *23*, 475702. [CrossRef] [PubMed]

44. Sader, J.E. Surface stress induced deflections of cantilever plates with applications to the atomic force microscope: Rectangular plates. *J. Appl. Phys.* **2001**, *89*, 2911–2921. [CrossRef]

45. Morin, P.; Raymond, D.; Benoit, P.; Maury, P.; Beneyton., R. A comparison of the mechanical stability of silicon nitride films deposited with various techniques. *Appl. Surf. Sci.* **2012**, *260*, 69–72. [CrossRef]

46. Colombi, P.; Bergese, P.; Bontempi, E.; Borgese, L.; Federici, S.; Keller, S.S.; Boisen, A.; Depero, L.E. Sensitive determination of the Young's modulus of thin films by polymeric microcantilevers. *Meas. Sci. Technol.* **2013**, *24*, 125603. [CrossRef]

 *coatings*

Article

Combining Thermal Spraying and Magnetron Sputtering for the Development of Ni/Ni-20Cr Thin Film Thermocouples for Plastic Flat Film Extrusion Processes

Wolfgang Tillmann [1], David Kokalj [1,*], Dominic Stangier [1], Volker Schöppner [2] and Hatice Malatyali [2]

[1]   Institute of Materials Engineering, TU Dortmund University, Leonhard-Euler-Straße 2, 44227 Dortmund, Germany; wolfgang.tillmann@tu-dortmund.de (W.T.); dominic.stangier@tu-dortmund.de (D.S.)
[2]   Kunststofftechnik Paderborn, Paderborn University, Warburger Straße 100, 33098 Paderborn, Germany; volker.schoeppner@ktp.uni-paderborn.de (V.S.); hatice.malatyali@ktp.uni-paderborn.de (H.M.)
*   Correspondence: david.kokalj@tu-dortmund.de; Tel.: +49-(0)231-755-4866

Received: 26 August 2019; Accepted: 20 September 2019; Published: 24 September 2019

Abstract: In the digitalization of production, temperature determination is playing an increasingly important role. Thermal spraying and magnetron sputtering were combined for the development of Ni/Ni-20Cr thin film thermocouples for plastic flat film extrusion processes. On the thermally sprayed insulation layer, AlN and BCN thin films were deposited and analyzed regarding their structural properties and the interaction between the plastic melt and the surfaces using Ball-on-Disc experiments and High-Pressure Capillary Rheometer. A modular tool, containing the deposited Ni/Ni-20Cr thin film thermocouple, was developed and analyzed in a real flat film extrusion process. When calibrating the thin film thermocouple, an accurate temperature determination of the flowing melt was achieved. Industrial type K sensors were used as reference. In addition, PP foils were produced without affecting the surface quality by using thin film thermocouples.

Keywords: nickel–chromium; thin film thermocouples; physical vapor deposition; flat film extrusion; foil quality

---

1. Introduction

Self-learning machines play a crucial role in today's and future production. In the 4th Generation Industrial Revolution (Industry 4.0), machines are digitally upgraded and merged to a big data environment [1]. The overall goal is the improvement of product quality and production scheduling by building up and using prediction tools [1]. However, self-learning machines are still far from being implemented in many industries [1]. Data required for these prediction tools can only be generated if the machines are equipped with corresponding sensors, such as temperature-, wear-, distance-, or pressure-sensors, that work exactly. In many areas, the determination of the temperature, in particular, plays an important role. For this purpose, especially thin film thermocouples are suitable, since they reveal a lower mass and thus a faster reaction time in contrast to bulk thermocouples [2]. Multi-functional PVD coatings are successfully used for the temperature measurement in combustion engines [3] or gas turbine engines [4], for example. However, there is also a need for thin film thermocouples in production processes. Therefore, different thin film thermocouples are currently developed for several applications, such as welding [5] or metal [6–9] and polymer processing [10,11]. However, the thin film thermocouples must be adapted to the respective application fields. On the one hand, the materials of the thermocouples need to be selected according to the desired temperature range. According to DIN EN 60584, eight different thermocouple pairs are available, whereas Fe–CuNi [12],

Cu–CuNi [13], NiCr–Ni [6–9] and Pt-10%Rh/Pt [14] are the most reported thin film thermocouples deposited by means of PVD. Apart from the standardized thermocouples, TiC–TaC [15], Al–Au [11], Ni–Cu [12,16], Ni–Fe [12], Cu–Fe [12], Chromel–Alumel [5,12], and Pt–Pd [4] thin film thermocouples are synthesized. On the other hand, depending on the application, the thin film thermocouples need to be protected against wear or oxidation processes by means of a suitable tribological cover layer. To enhance the wear resistance of thin film sensors, protection layers consisting of Al_2O_3 [7,14,16,17], AlN [17], or TiN [8,18] are mainly used. For electrical isolation purposes, HfO_2 [6] and Al_2O_3 [16] layers are utilized.

In plastic processing, thermoplastics are extruded into semi-finished products that can be processed further (e.g., films). The plastic is plasticized and formed in the die [19]. For a good product quality, the melt needs to flow through the die at constant pressure, throughput and temperature [20]. In particular, an accurate measurement of the melt temperature is preferred and helps to increase the productivity. Temperature sensors are part of every extruder system, so that the time stability and the absolute level of the melt temperature can be monitored. Malfunctions can be detected quickly and the reaction time to undesired temperature changes increases. Nevertheless, state-of-the-art touching sensors cannot be used in the die itself because of the negative impact of the sensors on the product quality. The measurement tip protrudes at different depths into the melt and can disturb the homogeneity in the process itself. Thin film thermocouples offer an alternative measuring method to conventional sensors due to their flat construction and low net weight. In addition, they offer the possibility to measure the temperature even when embedded in the wall. Especially, in the case of embedded sensors, it is important to measure the temperature of the melt and not of the die wall. Therefore, thin film thermocouples are to be thermally insulated from the die walls. Thermal sprayed barrier coatings, such as ZrO_2 and Al_2O_3, are particularly suitable for thermal insulation applications since they reveal pores and a higher thickness compared to sputtered coatings. Therefore, these coatings were applied by means of atmospheric plasma spraying and serve as substrate for the thin film thermocouples [19,20]. In particular, in film extrusion, it is important that the melt flow is not impaired to ensure a homogeneous film quality. Accordingly, the influence of different PVD top coatings on the wall friction and the adhesion of the melt is investigated. Since AlN coatings reveal high electrical resistance and high thermal conductivity, three AlN thin films deposited with different sputter parameters are selected. As an alternative, BCN is used as additional top coat. Finally, the application of the Ni/Ni-20Cr thin film thermocouple is proved in a real operation test und compared to a reference bulk thermocouple.

2. Materials and Methods

2.1. Deposition Process

For the deposition of the nitride thin films and the Ni/NiCr thin film thermocouples, the industrial scale magnetron coating system CC800/9Custom (CemeCon AG, Würselen, Germany) was utilized. The AlN and BCN thin films were synthesized on AISI 1045 steel substrates (Ø40 mm), which were prior coated with Al_2O_3 using atmospheric plasma spraying. Detailed spray parameters of the Al_2O_3 coating are reported in [21]. The nitride thin films were deposited in medium frequency (MF) mode using one B_4C (99.50% purity, Sindlhauser Materials GmbH, Kempten, Germany) and two Al (99.50% purity, CemeCon AG, Würselen, Germany) targets operating with 3150 W and 2× 1875 W at a total chamber pressure of 330 mPa. The frequency was set to 50 kHz with a duty cycle of 50%. Three AlN thin films and one BCN thin film were synthesized using different heating powers and Ar/N_2 gas flow rates according to Table 1. The heating power was changed to influence the crystallinity and the topography of the AlN thin films. Changing the Ar/N_2 gas flow ratio results in a change of the chemical composition of the coating. Therefore, it was possible to analyze which properties influence the melt flow, respectively, the friction behavior. The two-fold rotation system was biased with –150 V in MF mode during all deposition processes. The deposition time of all coatings was adjusted to achieve comparable thin film thicknesses of approximately 1200 nm. The thickness was measured analyzing the cross section of the coatings by means of scanning electron microscopy.

Table 1. Deposition parameters of the AlN and BCN thin films.

Coating	AlN-1	AlN-2	AlN-3	BCN-1
Heating power [W]	5000	7000	3000	7000
Ar/N$_2$ gas flow ratio	6.0	6.0	5.1	26.0

In addition to the reference nitride layers, Ni/Ni-Cr thin film thermocouples were synthesized for the temperature measurement of the plastic melt in flat film extrusion processes. Therefore, a modular tool system was developed. This system consisted of a screw-in bush (AISI 4140), a ceramic insert, and a locking nut. The bush can be integrated into the die by means of a drilling with a thread (M24 × 1.5). The ceramic insert was coated with the thin film thermocouples. By means of a step on the one side and a locking nut on the other side, the thin film thermocouple was positioned and fixed in the bush. The ceramic insert was composed of a steel core (AISI 1045), which was coated by atmospheric plasma spraying from all sides. For a sensitive temperature measurement, the insert had to be thermally insulated from the die walls, since the temperature of the on-rushing plastic melt was measured. The thermal insulation was achieved by a thermal sprayed multilayer design consisting of a 120-μm-thick NiCoCrAlTaY (AMDRY 997, Sulzer Metco, Pfäffikon, Switzerland) bond coat, a 530-μm-thick 7Y$_2$O$_3$-ZrO$_2$ (AMPERIT 817.7, HC Starck, München, Germany), and a 680-μm-thick Al$_2$O$_3$ (Metco 6062, Sulzer Metco, Pfäffikon, Switzerland) ceramic coating. After the coating processes, the Al$_2$O$_3$ coating was entirely ground to a thickness of 350 μm, resulting in a total coating thickness of 1000 μm. After the grinding process, the outsides were polished in several steps to a roughness of R_a = 0.212 ± 0.104 μm.

The step of the ceramic insert was manufactured using a micro milling machine HSPC 2522 (Kern, Eschenlohe, Germany) equipped with a solid carbide end mill, type 910 Marlin, with a corner diameter of 1 mm (Zecha, Königsbach-Stein, Germany). The milled surfaces reveal a roughness of R_a = 0.811 ± 0.116 μm.

After processing the final dimensions of the ceramic inserts, they were cleaned of contaminations by means of ethanol in an ultrasound bath. Thereafter, the Ni/Ni-Cr thin film thermocouple was deposited on the ceramic inserts using a steel masking system reported in [22] to synthesize the individual Ni and Ni-20Cr conducting paths. The conducting paths were sputtered in DC mode using a heating power of 1200 W and a Ar/Kr gas flow rate of 2.7 at a chamber pressure of 250 mPa. For the Ni conducting path, two Ni targets (99.99% purity, Sindlhauser Materials GmbH, Germany) were operated at 1875 W in DC mode, whereas for the Ni-20Cr path, a Cr target (99.95% purity, Sindlhauser Materials GmbH, Kempten, Germany) was operated at 1035 W, additionally. The deposition time of both conducting paths was adjusted to reveal similar thicknesses of 1050 ± 35 nm.

2.2. Characterization

The structural properties of the as-deposited nitride thin films were analyzed by means of X-Ray Diffraction (XRD) utilizing the diffractometer D8 Advance (Bruker, Madison, WI, USA). The thin films were investigated using Cr-radiation (2.29106 Å), whereas the current and voltage were set to 40 mA and 35 kV. To avoid an overlapping of the reflexes from the Al$_2$O$_3$ substrate, detector scans were performed, whereby the tube was set to an incident angle of 5°. The topography and morphology of the AlN and BCN thin films were investigated using the Scanning Electron Microscope (SEM) JSM 7001F (Jeol, Tokyo, Japan). The roughness R_a of the Al$_2$O$_3$ substrate, as well as the thin films, was measured by the confocal white-light microscope μSurf (NanoFocus, Germany). The mechanical properties hardness (H) and Young's modulus (E) were examined by means of a nanoindentation test using the nanoindenter G200 (Agilent Technology, Santa Clara, CA, USA). The evaluation was performed in a depth between 100 and 400 nm using the equations in accordance to Oliver and Pharr [23]. Thereby, a constant Poisson's ratio of 0.25 was assumed. The adhesion of the nitride thin films under plastic deformation was tested by a Rockwell indentation test in accordance with DIN EN ISO 26443 [24], using a force of 60 kgf and a dwell time of 4 s. Tribological investigations were carried out using a high-temperature Ball-on-Disc tribometer (CSM-Instruments, Peseux, Switzerland). The velocity was set to 0.1 m/s, the normal force to 10 N, and the radius to 8 mm with a total distance of 20 m per experiment. Counter body balls (Ø6 mm) made of Polypropylene (PP) were used, since this material was used for the operation test of the thin film thermocouples. The tribological investigations were carried out at 85 °C (Vicat Softening Temperature) and 153 °C (Heat Deflection Temperature) to simulate a non-steady-state operating point of the plastic flat film extrusion process during heating up or cooling down. The adhesion of the PP to the coating surface was investigated by means of SEM. Additionally, the wear of the PP balls was analyzed using a confocal light microscope type InfiniteFocus (Alicona, Raaba/Graz, Austria). By utilizing a High-Pressure Capillary Rheometer (HPCR), the influence of the different coatings on the melt flow was investigated. The measurement was carried out according to DIN 54811. When measuring with the HPCR, first, a purely thermal melting takes place in a pre-chamber. In order to melt the material completely, the definition of a residence time is important. For the investigations with the Moplen HP420 M, a duration of 4 min was specified. After melting, a piston pushed the melt through the tempered capillary. The capillary was preheated to a certain temperature by two thermal sensors. The operating temperature for the measurements was set to 230 and 250 °C. The capillary was preheated to the desired temperature by means of two sensors. The piston speed and the pressure loss of the melt flow were recorded simultaneously by means of a pressure sensor.

For the operational test, a wide-slot die was used, which allows to measure the melt temperature. The positioning of the thermocouples was important for the accuracy and reproducibility of measuring points. The thin film and industrial thermocouples were inserted near to the melt outlet from the die. The investigations were carried out using an extruder with a diameter of 45 mm and a typical three-zone screw for plasticizing (Battenfeld-Cincinnati Gmbh, Germany). The mass temperature measurements were carried out using a high density Polyethylen, while for the film production, a high flow polypropylene homopolymer (Moplen HP420M, lyondellbasell, Rotterdam, The Netherlands) was selected. Furthermore, for producing films, a chill roll unit was used (Collin GmbH, Maitenbeth, Germany). To compare the thin film sensors with conventional sensors of type K (Gneuss GmbH, Bad Oeynhausen, Germany), sensors with measuring tips (depth length) of 0, 0.5, and 1.5 mm were also utilized to measure the melt temperature and serve as reference. Besides the operational tests, the quality of the produced film was examined for evaluating the influence of the measuring tip of the industrial sensors. The optical properties of the film were determined by measuring the reflection and the transmission. An optical spectroscope HR2000+ (Ocean optics, EW Duiven, The Netherlands) enabled the measurement of the film quality. The measuring setup was in accordance with DIN 5036-3. Optical spectroscopy enabled a fast quality criterion for the evaluation of samples. In this method, a light beam with an intensity I_0 irradiated the sample, while the exit intensity was recorded. The radiation sources covered a selected wavelength range, respectively, a certain spectrum. The intensity of the light depends on the wavelength λ and, accordingly, a function $I(\lambda)$ of the light intensity [25]. The spectrum of the incident intensity was compared over the wavelength with the spectrum of the incident intensity. In the present study, the visual spectrum (VIS), which reveals wavelengths of approximately 380–780 nm, was used. When the light beam hits the matter, the light is distributed to different parts and the following output intensities are generated: The reflected I_R, the transmitted I_T, the scattered I_S, and the absorbed intensity I_A. These intensities form in sum I_0 [25].

3. Results

3.1. AlN/BCN Thin Films

3.1.1. Structure and Morphology

XRD patterns of the three AlN and the one BCN coatings synthesized with different deposition parameters are presented in Figure 1. For the AlN thin films, the positions of the reflections are $2\theta \approx$ 50.2°, $2\theta \approx$ 54.7°, $2\theta \approx$ 57.7°, $2\theta \approx$ 77.4°, $2\theta \approx$ 94.8°, and $2\theta \approx$ 106.6°, which correspond to the hexagonal AlN phase (JCPDS 25-1133). The reflections of the three AlN thin films reveal comparable intensities, showing a similar crystallinity. Nevertheless, slight differences can be detected. The AlN-3 thin film shows the highest crystallinity, which is favored by the higher nitrogen flow during the coating process. At constant gas flow, the AlN-2 film shows a slightly higher crystallinity compared to the AlN-1 film, which is due to the higher deposition temperature. Compared to the AlN-3 thin film, the AlN-1 und AlN-2 thin films reveal small additional peaks at $2\theta \approx$ 69.0° and $2\theta \approx$ 106.4°, which correspond to the cubic AlN phase (JCPDS 46-1200). However, none of the AlN thin films show a high crystallinity, which is also shown by the broad peaks. Some of these also overlap with the comparable structure of the substrate, which makes accurate evaluation difficult. The wurtzite-type AlN was already obtained by Aissa et al. for a dc and high power impulse magnetron sputtering process [26]. The remaining reflections, marked with a black circle, belong to the cubic (α) and trigonal (γ) Al_2O_3 phase of the thermally sprayed substrate. Concerning the BCN coating, basically, only peaks of the substrate are visible, which indicates a nearly amorphous state of the thin film. Only slight reflections at $2\theta \approx$ 52.8° and $2\theta \approx$ 57.4° are observed, which coincide with the position of rhombohedral B_4C.

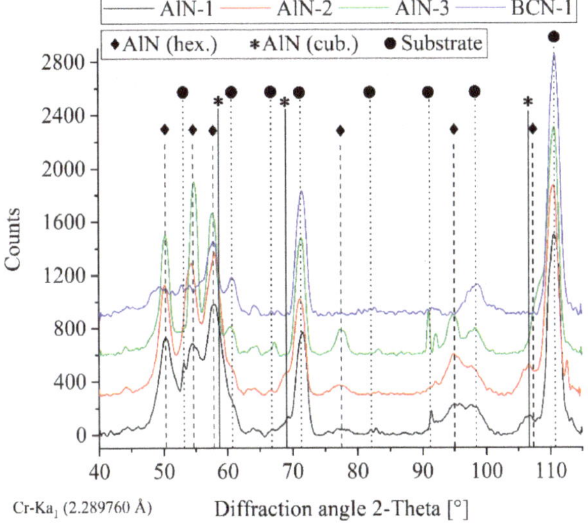

Figure 1. XRD patterns of the AlN-1, AlN-2, AlN-3, and BCN thin films deposited with various parameters.

Additionally, the nitride thin films were analyzed regarding topography and morphology by means of SEM, as shown in Figure 2. It can be seen that the thermally sprayed Al_2O_3 coating, which was used as substrate for the thin films, reveals a fine topography accompanied by a small crack pattern in the polished state. The AlN-1 and AlN-2 thin films possess a coarse cauliflower-like topography. By contrast, the AlN-3 thin film shows a fine structured topography. Accordingly, a high Ar/N_2 gas flow ratio of 6.0 and high heating powers of 5000 and 7000 W, as is the case for the AlN-1 and AlN-2 coatings, lead to a coarse topography. The structure of the topography can be correlated with the XRD pattern shown in Figure 1. The AlN-1 and AlN-2 thin films reveal a lower crystallinity compared to the AlN-3 thin film and a small amount of a second phase, which results in the coarser growth behavior. The BCN-1 coating reveals an even finer structure, which is comparable to the polished Al_2O_3 surface. The cross-sections of all thin films show a glass-like featureless structure.

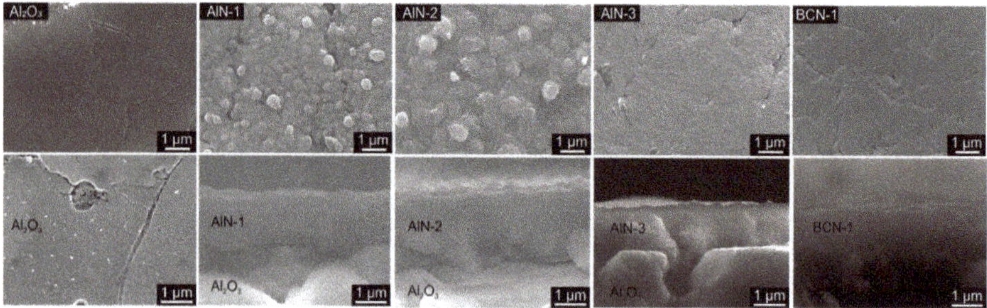

Figure 2. SEM images of the topography (top) and morphology (bottom) of the AlN-1, AlN-2, AlN-3, and BCN thin films, as well as the thermally sprayed Al_2O_3 coating.

3.1.2. Mechanical Properties

Using nanoindentation, the mechanical properties—hardness and Young's modulus—of the thin films, as well as the Al_2O_3 coating, were analyzed. As listed in Table 2, the Al_2O_3 coating reveals a hardness of 13.8 ± 3.9 GPa and a Young's modulus of 155.2 ± 31.9 GPa. The AlN-1 and AlN-2 thin films reveal the lowest hardness of 6.0 ± 1.6 GPa and 6.2 ± 2.0 GPa, respectively. A higher hardness of 13.6 ± 3.0 GPa is analyzed for the AlN-3 thin films, which reveals a finer structure compared to the AlN-1 and AlN-2 thin films. The highest hardness of 19.1 ± 2.3 GPa is obtained for the BCN-1 coating. With increasing hardness, the Young's modulus increases from 98.6 ± 21.2 GPa (AlN-1) to 187.7 ± 20.2 GPa (BCN-1). AlN thin films deposited on stainless steel substrates by means of magnetron sputtering reveal similar hardness values between 6 and 14 GPa in dependence on the bias-voltage as reported by Choudhary et al. [27]. The H/E ratio, which demonstrates the resistance to plastic deformation, is between 0.048 (AlN-2) and 0.102 (BCN-1). Since a higher value demonstrates a higher resistance to plastic deformation, the BCN-1 and AlN-3 thin films, as well as the uncoated Al_2O_3 coating, are the best choices as top layer for the application in plastic melt extrusion dies.

Table 2. Roughness and mechanical properties of the nitride thin films, as well as the Al_2O_3 coating.

Coating	Al_2O_3	AlN-1	AlN-2	AlN-3	BCN-1
Roughness Ra [μm]	0.21 ± 0.10	0.36 ± 0.06	0.30 ± 0.08	0.26 ± 0.06	0.35 ± 0.08
Hardness [GPa]	13.8 ± 3.9	6.0 ± 1.6	6.2 ± 2.0	13.6 ± 3.0	19.1 ± 2.3
Young's modulus [GPa]	155.2 ± 31.9	98.6 ± 21.2	130.4 ± 28.5	167.5 ± 24.1	187.7 ± 20.2
H/E	0.089	0.061	0.048	0.081	0.102

Hence, the thin films were exposed to high pressures in the nozzles without abrasive particles occurring. The adhesion of the thin films was analyzed by means of a Rockwell indentation test, which represents the thin film behavior under plastic deformation. The corresponding light microscope images of the indents are shown in Figure 3. According to DIN EN ISO 26443, the adhesion of the thin films can be classified into four different classes. Thereby, class 0 represents a coating without cracks and local delamination, and class 3, full delamination of the coating after indentation. The thin films AlN-2 and BCN-1 are classified as class 1, since they reveal only slight cracks around the indent. AlN-1 and AlN-3 correspond to class 2, showing local delamination. Accordingly, the adhesion strength of the thin films decreases with decreasing heating power, respectively, deposition temperature. The lowest adhesion is shown by the AlN-3 thin film, deposited at the lowest temperature and with the highest nitrogen gas flow.

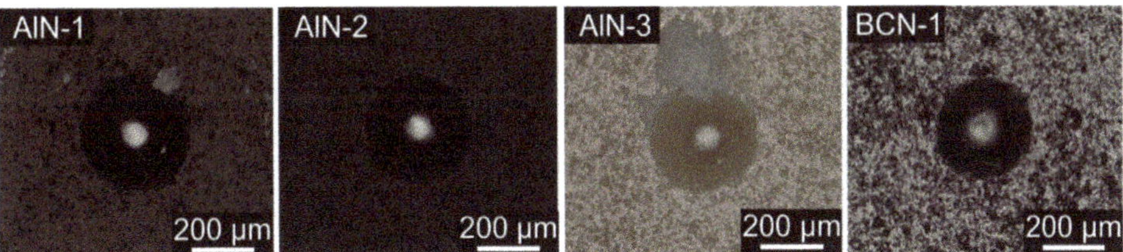

Figure 3. Light microscope images of the Rockwell indents of the AlN-1, AlN-2, AlN-3, and BCN thin films.

3.1.3. Tribological Properties

Polymer processing mainly TiN, CrN, TiAlN, $Cr_{1-x}Al_xN$, CrAlON, and DLC coatings deposited by means of PVD were investigated and used [28–30]. It was shown that the addition of Al to CrN increases the contact angle between the surface and Polycarbonate melt, resulting in a lower adhesion. Therefore, the AlN and BCN thin films were investigated regarding their tribological properties. Tribological experiments were conducted at 85 and 153 °C using PP counter balls, whereby the temperatures are the Vicat Softening Temperature and the Heat Deflection Temperature. These temperatures were selected to simulate the adhesion behavior of the PP melt to the different top layer-coated thin film thermocouples during heating up or cooling down processes of the thin film extrusion process. Different friction coefficients between the melt and the thin film thermocouple, as well as the uncoated extrusion die, can lead to inhomogeneous film qualities due to different friction forces. Since the extrusion die was made of AISI H11 (1.2343) steel, this steel was used as reference for the tribological experiments. In Figure 4, the friction coefficients of the AlN-1, AlN-2, AlN-3, and BCN-1 thin films, as well as the steel substrate and the Al_2O_3 coating, are shown for temperatures of 85 and 153 °C using PP counter balls. It was observed that there is no temperature dependency of the friction coefficient, except for the reference steel and the Al_2O_3 coating. At 85 °C, a friction coefficient of 0.52 ± 0.11 is measured for the steel substrate, which drops down to 0.27 ± 0.05 at 153 °C. A contrary behavior is observed for the Al_2O_3 coating; the friction coefficient increases from 0.42 ± 0.10 at 85 °C to 0.66 ± 0.16 at 153 °C. Concerning the thin films, they basically reveal comparable friction coefficients independent from the used deposition parameters. The lowest friction coefficient of 0.28 ± 0.05 is obtained for the AlN-2 thin film, which slightly increases in the order AlN-1, AlN-3 up to 0.39 ± 0.07 for the BCN-1 thin film at 85 °C. In a different study, Ball-on-Disc experiments were also used to investigate the friction coefficient between CrAlN, respectively, DLC thin films and Polycarbonate counter balls [31]. It was shown that the friction coefficient depends on the temperature, and the adhesion behavior is related to the surface energies of the deposited thin films and surfaces. Moreover, it was reported that the friction coefficient depends on the roughness of the surfaces. Smoother surfaces lead to a higher fraction of the adhesive effect on the friction, which in turn leads to a higher coefficient of friction [32]. In fact, the obtained friction coefficients at 85 °C can be tendentially correlated with the roughness. The polished steel surface reveals the smoothest surface with a roughness of R_a = 0.006 ± 0.001 μm and the highest friction coefficient. The second lowest roughness is shown by the thermally sprayed Al_2O_3 coating, which shows the second highest roughness. Higher roughness and lower friction coefficients were analyzed for the thin films. At 153 °C, the influence of the roughness on the friction coefficient is not observed. The melting point of PP is between 160 and 165 °C, and therefore the mechanical properties of PP are already decreased at 153 °C. The surface of the Al_2O_3 coating contains defects such as pores, as shown in Figure 2, which affect the friction behavior, especially in contact with a soft material as also shown by the large deviation bar of the friction coefficient. Therefore, the Al_2O_3 coating reveals a higher friction coefficient at 153 °C compared to the other thin films.

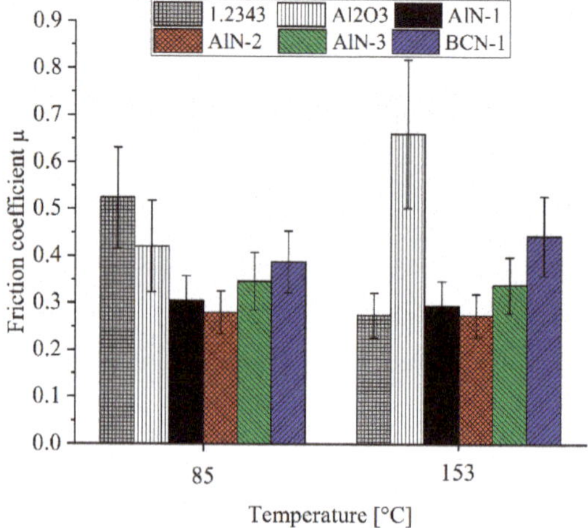

Figure 4. Friction coefficient of the AlN-1, AlN-2, AlN-3, and BCN-1 thin films, as well as the steel substrate and the Al$_2$O$_3$ coating, at 85 and 153 °C using PP counter balls.

In addition to the frictional forces, a comparable adhesion of the plastic to the thin film thermocouple and the extrusion die is to be ensured. Therefore, the wear of the PP balls after the Ball-on-Disc experiments was analyzed, which is an indirect hint for the adhesion behavior to the used coatings. As visualized in Figure 5, the wear coefficients of the PP balls are between $16.9 \pm 1.3 \times 10^{-5}$ mm^3/Nm (AlN-3) and $34.3 \pm 5.1 \times 10^{-5}$ mm^3/Nm (AlN-1) for the PVD coated counter parts, whereas a wear coefficient of $26.7 \pm 3.4 \times 10^{-5}$ mm^3/Nm is analyzed for the uncoated steel counterpart at 85 °C. Using the thermally sprayed Al$_2$O$_3$ coating as a counter body, a wear coefficient of $10.6 \pm 0.6 \times 10^{-5}$ mm^3/Nm is noticed. Accordingly, at 85 °C, all counterparts cause a comparable wear coefficient of the PP balls, since the maximum difference of the wear coefficient is 23.7×10^{-5} mm^3/Nm within the analyzed counter parts.

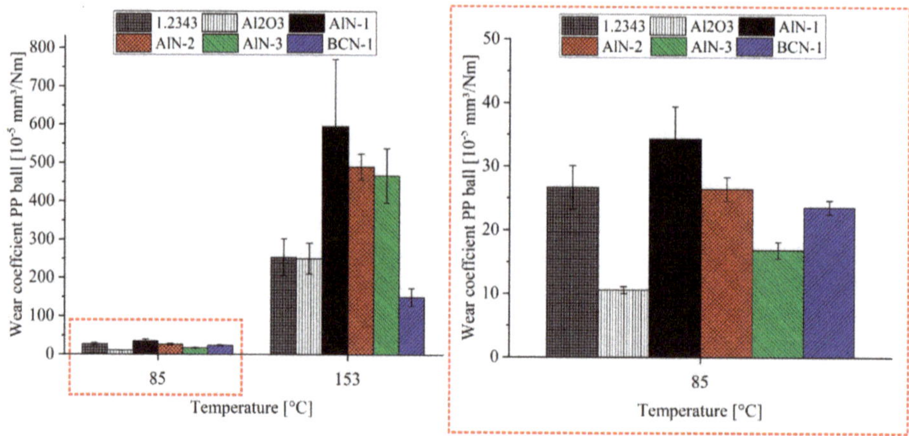

Figure 5. Wear coefficient of the PP counter balls after the sliding against the AlN-1, AlN-2, AlN-3, and BCN-1 thin films, as well as the steel substrate and the Al$_2$O$_3$ coating, at 85 and 153 °C.

At 153 °C, an increased wear coefficient of the PP balls is detected, compared to the tribological tests at 85 °C. The wear coefficient of the PP balls is increased by the factor 10 to 20, since the mechanical properties of the PP balls are reduced at elevated temperatures. Analyzing the wear coefficient at 153 °C, the differences between the counter parts become more visible compared to 85 °C. The PP balls slid against the steel surface reveal a wear coefficient of $253.7 \pm 48.7 \times 10^{-5}$ mm^3/Nm, which is comparable to a wear coefficient of $250.3 \pm 41.0 \times 10^{-5}$ mm^3/Nm for the Al$_2$O$_3$ coating. A lower coefficient of $149.9 \pm 22.9 \times 10^{-5}$ mm^3/Nm is obtained for the BCN-1 counterpart, and higher wear coefficients between $467.0 \pm 70.5 \times 10^{-5}$ mm^3/Nm (AlN-3) and $594.8 \pm 175.2 \times 10^{-5}$ mm^3/Nm (AlN-1) for the AlN thin films. The used PP counter ball is a non-polar plastic, revealing a low fraction of polar surface energy. Accordingly, only the disperse fraction of the free surface energy influences the adhesion behavior to the PP counterpart [33]. Theiss et al. reported that the work of adhesion to PP, which is calculated based on the surface energies, is higher for AlN-rich thin films compared to steel [34]. Especially, for thin films grown in the hexagonal structure, a high work of adhesion is observed [34]. Accordingly, the higher wear of the PP balls slid against the AlN thin films compared to the steel counter body can be related to the surface energies. Moreover, this mechanism is overlaid with the surface roughness. As listed in Table 2, the used deposition parameters of the AlN thin films lead to a change in the roughness. The roughness Ra varies between 0.26 ± 0.06 μm (AlN-3) and 0.36 ± 0.06 μm (AlN-1). The wear of the balls decreases with decreasing roughness in the order AlN-1, AlN-2, AlN-3.

Summarized, within the AlN thin films, the deposition parameters influence the friction coefficient and the wear coefficient of the PP balls at 85 °C, as well as 153 °C. At 153 °C, the influence of the deposition parameters of the AlN thin films on the friction coefficient is lower compared to the wear coefficient. Concerning the wear coefficient at 153 °C, the wear of the PP balls is more influenced by the counter body material than the used deposition parameters of a coating, since the difference between the different materials is higher than the difference within the AlN thin films.

Corresponding SEM images of the adhesions from the PP balls to the different coatings after the test at 85 °C are shown in Figure 6. No adhesion of PP to the surface of the uncoated 1.2343 substrate can be observed. In addition, no adhesions to the Al$_2$O$_3$, AlN-3, and BCN-1 surfaces have occurred. In contrast to that, local PP accumulations are observed on the AlN-1 and AlN-2 thin films, whereas the amount adhered to the AlN-1 thin film is slightly higher. On the one hand, the adhesion to these thin films can be explained by the cauliflower-like surface, resulting in a higher surface roughness in contrast to the other coatings, as shown in Figure 2. On the other hand, the adhesive layer is only observed for the AlN thin films featuring a small fraction of a second phase, as discussed in Section 3.1.1.

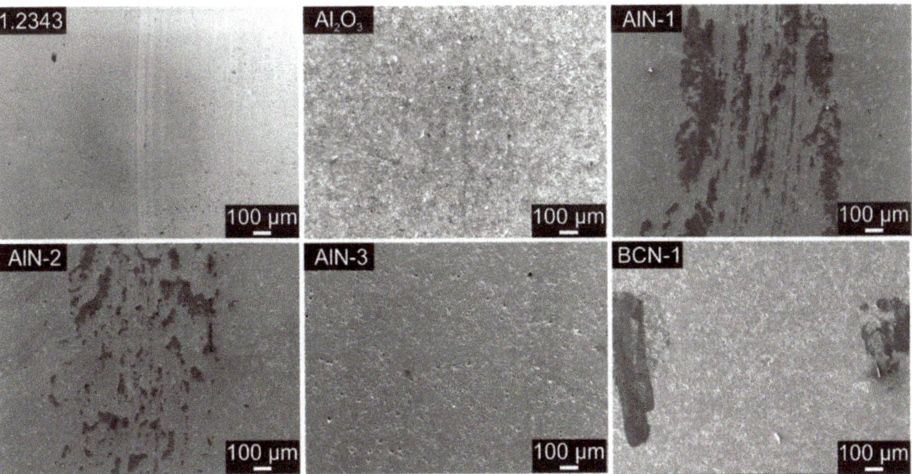

Figure 6. SEM images of the different surfaces after the Ball-on-Disc experiments at 85 °C.

SEM images of the different surfaces after the Ball-on-Disc experiments at 153 °C are visualized in Figure 7. In contrast to the surfaces analyzed at 85 °C, at 153 °C for all coatings, no rubbing of the PP balls on the surfaces can be noted, which could be related to the change in properties of the PP material with increasing temperature. At 85 and 153 °C, there is no general correlation between the ball wear coefficients and the formation of an adhesive layer. However, the formation of the adhesive layer on the AlN-1 and AlN-2 thin films at 85 °C correlates with the highest ball wear rates within the three AlN thin films.

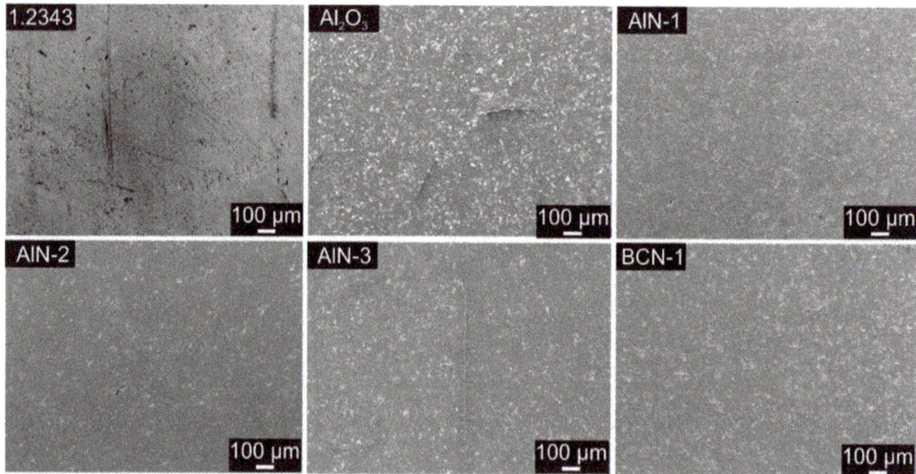

Figure 7. SEM images of the different surfaces after the Ball-on-Disc experiments at 153 °C.

Furthermore, the influence of the different coating surfaces on the melt flow was investigated at higher temperatures corresponding to the extrusion process. Therefore, the coatings were analyzed in a High-Pressure Capillary Rheometer (HPCR). By means of coated surfaces of specially designed exchangeable inserts, the influence on the flow properties of the melt was investigated. Figure 8 shows the basic setup of the capillary and the coated inserts on the example of the AlN-3 thin film.

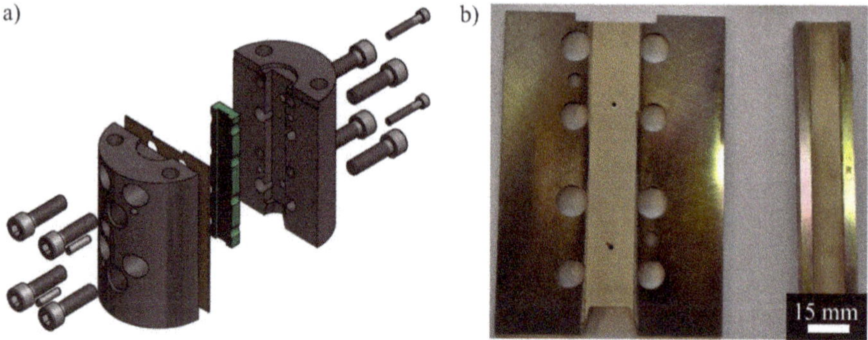

Figure 8. (**a**) Setup High-Pressure Capillary Rheometer and (**b**) coated exchangeable inserts with AlN-3.

Flow anomalies, which are due to the different surface qualities based on the compositions of the coatings, can be detected with a so called "critical" wall shear stress. Until the critical wall shear stress is reached, the melt flow behavior is like "wall sticking". Therefore, flow anomalies cannot be detected. The onset of the throughput jump depends on, among other things, the melt temperature and the polymer type. In the transition area above the critical wall shear stress, the throughput multiplies many times over, while the wall shear stress remains constant. This range is also referred to as the limiting wall shear stress. Above a certain shear rate, the wall shear stress increases again and the effect of sliding occurs.

Figure 9 shows for the two operating temperatures of 230 and 250 °C and the obtained values of the shear rate in dependence of the shear stress using HPCR. Both diagrams show a steady increase of the shear rate over the wall shear stress. A comparison of the coated measurement curves with the reference measurement (grey curve with squares) shows no significant differences. Minimal deviations between the curves are found and can be attributed to the used setup. From the obtained values, it is concluded that no particular flow anomalies can be detected with the surface coatings present. A critical wall shear stress cannot be detected at the typical shear rates in the extrusion process $\left(\dot{\gamma} = 10^0 - 10^{-4}\frac{1}{s}\right)$. Consequently, all investigated coatings can be used as top layers for the thermocouples, since there is no difference in the melt flow behavior between the uncoated and coated surfaces of the wide-slot nozzle.

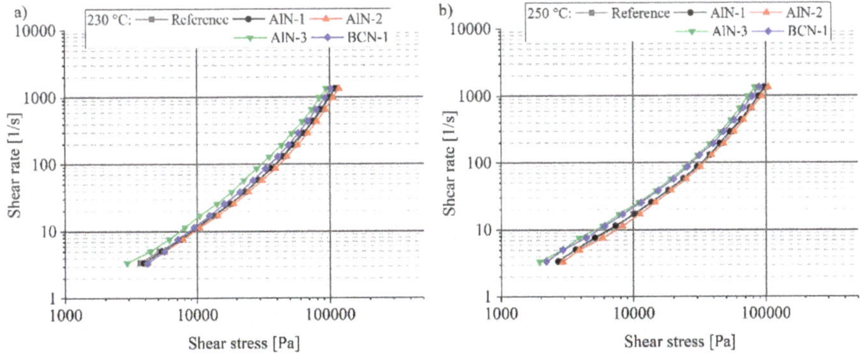

Figure 9. Critical wall shear stress at (**a**) 230 °C and (**b**) 250 °C, independent of the thin film.

3.2. Insertable Thin Film Thermocouple

The insertable thin film thermocouple is designed to be exchangeable, applicable to plastic film extrusion dies. As shown in Figure 10, the unity consists of a bush, which can be screwed into any die revealing an appropriate drilling (M24). The ceramic insert contains the thin film thermocouple and can be inserted into the bush, and is fixed by a step on the bottom side and a cap on the top side. The total length of the screw-in bush, including the top cap, is 63.5 mm.

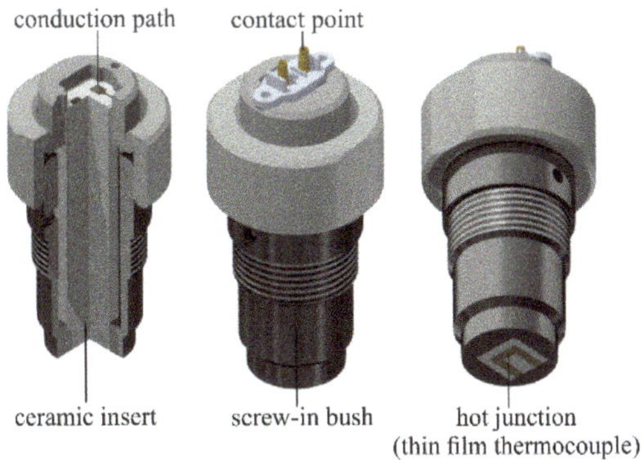

Figure 10. Structure of the exchangeable thin film thermocouple.

The Ni and Ni-20Cr conducting paths (width of 0.5 mm), deposited on the ceramic insert using a masking system, are shown in Figure 11. On the bottom side (step), the conducting paths are connected to form the hot junction length of 1.5 mm. The conducting paths run over the lateral surface to the top side. On the top side, the conducting paths end with contact points (2.5 × 2.5 mm^2), as visualized in Figure 10. Thermocouple balancing lines can be connected to these points, either by brazing or spring probe pins, to conduct the generated thermovoltage to the measuring system.

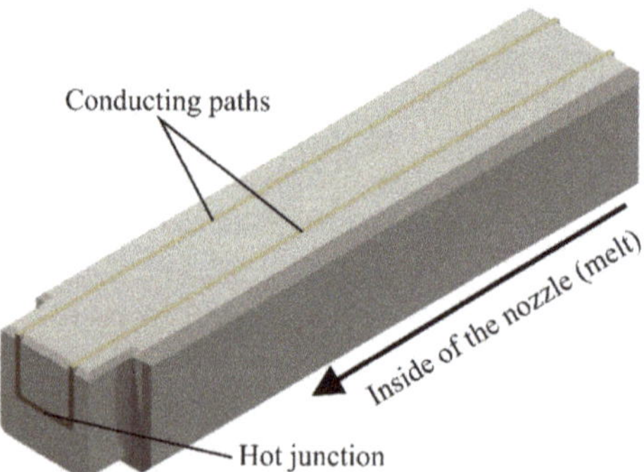

Figure 11. Ceramic insert showing the conducting paths of the Ni/Ni-20Cr thin film thermocouple.

The multilayer design of the thermally coated insert is shown in Figure 12. Primarily, the steel core is coated with a bond coat (NiCoCrAlTaY) to enhance the adhesion of the ceramic layers to the steel substrate, especially at higher temperatures. The bond coat is followed by a coarse $7Y_2O_3$-ZrO_2 coating to ensure the thermal insulation between the thin film thermocouple and the die, as well as the steel core of the insert, for an accurate measurement of the temperature of the plastic melt variations. Especially for wall mounted thermocouples, it is reported that the measurement of temperature fluctuations is difficult when the thermal insulation is not adequate [10]. The $7Y_2O_3$-ZrO_2 coating is layered by a dense Al_2O_3 coating, which improves the adhesion of the thin film thermocouples due to a lower content of defects, such as pores. Moreover, Al_2O_3 reveals a higher electrical resistance compared to ZrO_2 to guarantee no electrical short of the thermocouple.

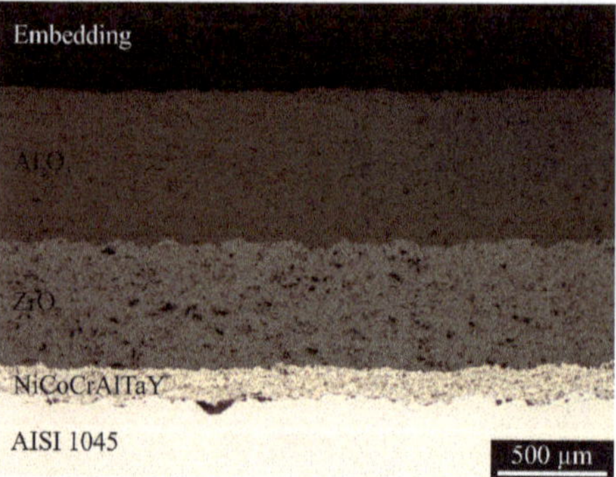

Figure 12. Cross-section of the multilayer design of the thermally sprayed insulating coatings of the ceramic insert.

The extrusion nozzle is made of steel and, therefore, the inserted tool with the thin film thermocouple should reveal the same tribological properties. As shown by the Ball-on-Disc experiments, all coatings reveal no adhesions of the PP counter ball on the surface at 153 °C. Concerning the wear of the PP ball, similar values are obtained for the steel surface and the thermally sprayed Al_2O_3 coating. High-Pressure Capillary Rheometer experiments also demonstrate no significant differences in the melt flow for the different coatings. Therefore, the thin film thermocouple was not covered by an additional nitride layer for the operation test.

3.3. Operation Test

3.3.1. Validation of the Temperature Measurement

Before the operational testing, the calibration curves for the thermocouples have to determined. In a previous study [20], the procedure was described and the dependence of the thermovoltage and temperature for the thermocouples was found. In the present study, the mass temperature was measured shortly before the outlet of the die. The used experimental setup is shown in Figure 13.

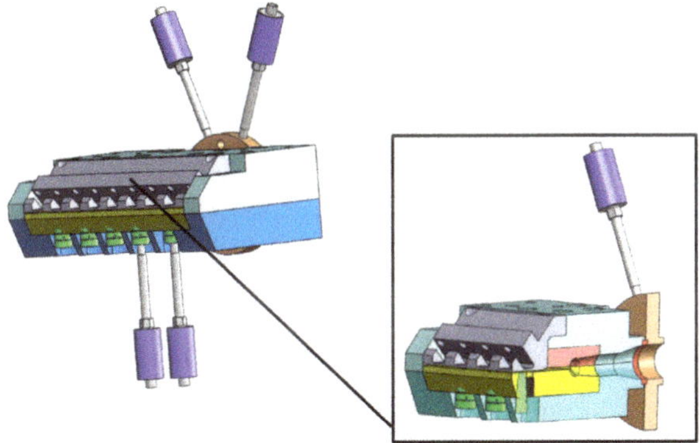

Figure 13. Experimental setup showing the array of the thermocouples in the wide-slit nozzle.

After the extrusion line reached a steady state for the cylinder heating and the used material, the embedded thin film thermocouple (TE) was inserted in the die and the recording of the measurement signal began. All four elements were inserted at the same time. Figure 14 shows the heating process of the thermocouple (grey line) until a static state was reached. All other temperature sensors were installed in advance to check the tool temperature. In total, the thermocouple required 130 s until 90% of the final value was reached, and approximately 1000 s until the stationary state was reached.

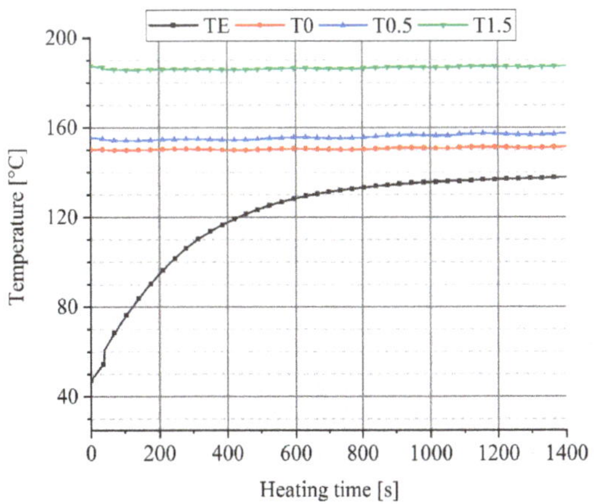

Figure 14. Heating curve of the thin film thermocouple in the plastic flat film extrusion process.

After the steady state of the thermocouple was reached, the measurement with all sensors was recorded over a time of 1.5 h. The measurement of the melt temperature at the mold outlet was additionally determined in 5 min intervals by means of a thermocouple as a guideline value. The surface temperature was also measured with a laser to verify the values. Figure 15 shows the temperature curve over time.

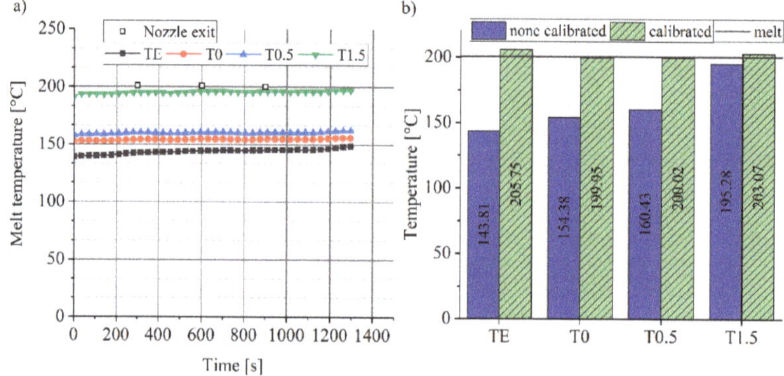

Figure 15. (**a**) Monitor conditioning of the melt temperature and (**b**) influence of the calibration on the measured mass temperature.

During the measuring process, no temperature peaks were detected (Figure 15a). All sensors displayed a constant measuring signal within the time interval. Due to the measuring tip of the T1.5, the sensor was placed deeper in the melt and measured the melt temperature significantly better. These measured values and the mathematical equations for the calibration curves were used to adjust accuracy of the measured melt temperatures. Figure 15b shows the adjusted temperatures. All thermocouples show relatively similar measured values after calibration. Compared to the mass temperature of 200.6 °C obtained by a laser at the nozzle exit, the sensors T0 and T0.5 measure with the lowest deviation after calibration. The deviation of the thin film thermocouple is 2.52%. Especially in polymer extrusion, the determination of the temperature is important, since even small deviations lead to a high change of the shear viscosity of the melt [35]. However, all used thermocouples, including the thin film thermocouple, can be used for the temperature monitoring of the mass in the extrusion die process after recalibration regarding the mounting place in the nozzle. Before the recalibration, the accuracy of the measured mass temperature increased with extent into the melt. For obtaining die melt temperature profiles in extrusion processes, generally, thermocouple mesh arrangements are inserted in the melt flow [10]. However, this technique can only be used in test conditions, since the melt flow is disturbed. Using many thin film thermocouples, a temperature profile can be obtained without affecting the melt flow.

3.3.2. Foil Quality

The optical film properties were determined by measuring gloss and transmission. Ten samples (dimensions 50×100 mm^2) were taken for evaluation during flat film production. The transmission and the gloss of the films were tested based on the described experimental setup. For the transmission, the determination of the light intensity I_0 is necessary as reference. The transmitted light intensity I_T was measured by clamping the samples in the sample holder. The difference between I_0 and I_T is the transmission spectrum. For the reflection, a white comparison sample is required, which completely reflects the entire light intensity I_0. Ocean Optics provided white specimens for the comparison sample. Then, the samples were irradiated with the light intensity I_0 at the angle of 2° defined in DIN 5036-3, and the reflected part I_R was measured.

Table 3 shows the reflection results for settings 1 and 2 as reflectance [%]. The velocity of the trigger unit was changed between the two settings. While at setting 1, the velocity reaches 1.5 mm/s, at setting 2 the velocity is 1.7 mm/s. For both settings, the values are between 29.91% and 31.88% and, accordingly, there is no significant difference between the different sensors. The wall-flush thermal sensors and the thin film thermocouples have a slightly lower value than the others.

Table 3. Results of the reflection results in dependence of the used thermocouple.

Thermocouple	Setting 1	Setting 2
	Reflectance [%]	
Reference	31.387	31.456
T0	30.937	31.228
T0.5	31.485	30.876
T1.5	31.087	31.150
Thin film thermocouple	30.964	29.910

Table 4 shows the transmission results for settings 1 and 2 as transmittance [%]. There are no significant differences in the settings. Therefore, the geometry of the thermocouples does not influence the transmittance. Only for setting 2, the measured values are 5% higher than setting 1. This effect is to be explained due to the fact that setting 2 produces thinner films, and an increased light intensity penetrates the sample.

Table 4. Results of the transmission results in dependence of the used thermocouple.

Thermocouple	Setting 1	Setting 2
	Transmittance [%]	
Reference	77.47	85.03
T0	80.33	85.00
T0.5	79.09	85.32
T1.5	78.90	84.75
Thin film thermocouple	79.39	85.52

4. Conclusions

In this study, AlN and BCN thin films were magnetron sputtered on thermally sprayed Al_2O_3 coatings using different deposition parameters. In the case of the AlN thin films, a high Ar/N_2 gas flow ratio of 6.0 and a high heating power led to a coarse structure with a reduced hardness from about 14 to 6 GPa. Since these films are proposed to be used in flat film extrusion nozzles, tribological experiments were conducted. Using a high deposition temperature and high Ar/N_2 gas flow ratio for the AlN coatings favor the adhesion of Polypropylen to the thin film surfaces after Ball-on-disc experiments at 85 °C. The formation of the adhesive layer is caused by the higher roughness of the AlN thin films, which can be related to the physical structure, and the work of adhesion. In contrast, no adhesions to the steel surface, the thermally sprayed Al_2O_3 coating, the BCN thin film, or a AlN thin deposited with lower temperature and higher Ar/N_2 gas flow ratio were observed. At 153 °C, for all coatings, no adhesions emerged, since the mechanical properties of the PP ball were changed with the temperature. The interaction of the thin film thermocouples with the plastic melt was investigated in application tests. Compared to the uncoated tool surface, the thin films reveal no negative effects on the wall shear stress and the wall shear speed. Accordingly, an effect of the thermocouple on the foil quality could not be detected in the samples examined under the spectroscope. Thus, flat films with constant reflection properties could be produced by measuring the melt temperature using thin film thermocouples.

Ni/Ni-20Cr thin film thermocouples were deposited on newly developed exchangeable tool inserts and deployed in a flat film extrusion process. The measurement signal of the thin film thermocouple and industrial reference thermocouples were tested using a specially designed die. No temperature peaks were detected during the measuring process, demonstrating a stable measurement behavior. At a melt temperature value of 200.57 °C, the obtained value by the help of the thin film thermocouple was 205.75 °C, which is slightly higher compared to the industrial sensors. The deviation of the thin film thermocouple did not exceed 2.52%, which is within a typical tolerance range of a type K sensor. Within the proposed attempt, PVD thermocouples are a promising approach which can be used for online monitoring. However, the used design can be used to further enhance the performance of the wide-slit nozzle by using wear resistant top layers.

Author Contributions: Conceptualization, D.K., H.M. and D.S.; Methodology, D.K. and H.M.; Validation, D.K., H.M. and D.S.; Investigation, D.K. and H.M.; Writing—Original Draft Preparation, D.K. and H.M.; Writing—Review and Editing, W.T., D.S. and V.S.; Supervision, W.T. and V.S.; Funding Acquisition, W.T. and V.S.

Funding: The authors gratefully acknowledge the financial support of project 18627 N of the German Federation of Industrial Research Associations (AiF) and the Forschungskuratorium Maschinenbau e.V. (FKM).

Acknowledgments: We acknowledge financial support by Deutsche Forschungsgemeinschaft and Technische Universität Dortmund/TU Dortmund University within the funding programme Open Access Publishing. In addition, the authors thank Dirk Biermann and Eugen Krebs from the Institute of Machining Technology (TU Dortmund University) for machining the ceramic inserts and providing the confocal 3D microscope.

References

1. Lee, J.; Kao, H.-A.; Yang, S. Service innovation and smart analytics for industry 4.0 and big data environment. *Procedia CIRP* **2014**, *16*, 3–8. [CrossRef]
2. Yust, M.; Kreider, K.G. Transparent thin film thermocouple. *Thin Solid Films* **1989**, *176*, 73–78. [CrossRef]
3. Marr, M.A.; Wallace, J.S.; Chandra, S.; Pershin, L.; Mostaghimi, J. A fast response thermocouple for internal combustion engine surface temperature measurements. *Exp. Therm. Fluid Sci.* **2010**, *34*, 183–189. [CrossRef]
4. Tougas, I.M.; Gregory, O.J. Thin film platinum–palladium thermocouples for gas turbine engine applications. *Thin Solid Films* **2013**, *539*, 345–349. [CrossRef]
5. Zhao, J.; Li, H.; Choi, H.; Cai, W.; Abell, J.A.; Li, X. Insertable thin film thermocouples for in situ transient temperature monitoring in ultrasonic metal welding of battery tabs. *J. Manuf. Process.* **2013**, *15*, 136–140. [CrossRef]
6. Basti, A.; Obikawa, T.; Shinozuka, J. Tools with built-in thin film thermocouple sensors for monitoring cutting temperature. *Int. J. Mach. Tools Manuf.* **2007**, *47*, 793–798. [CrossRef]
7. Biermann, D.; Kirschner, M.; Pantke, K.; Tillmann, W.; Herper, J. New coating systems for temperature monitoring in turning processes. *Surf. Coat. Technol.* **2013**, *215*, 376–380. [CrossRef]
8. Shinozuka, J.; Basti, A.; Obikawa, T. Development of cutting tool with built-in thin film thermocouples for measuring high temperature fields in metal cutting processes. *J. Manuf. Sci. Eng.* **2008**, *130*, 34501. [CrossRef]
9. Tillmann, W.; Vogli, E.; Herper, J.; Biermann, D.; Pantke, K. Development of temperature sensor thin films to monitor turning processes. *J. Mater. Process. Technol.* **2010**, *210*, 819–823. [CrossRef]
10. Abeykoon, C.; Martin, P.J.; Kelly, A.L.; Brown, E.C. A review and evaluation of melt temperature sensors for polymer extrusion. *Sensors Actuators A Phys.* **2012**, *182*, 16–27. [CrossRef]
11. Debey, D.; Bluhm, R.; Habets, N.; Kurz, H. Fabrication of planar thermocouples for real-time measurements of temperature profiles in polymer melts. *Sensors Actuators A Phys.* **1997**, *58*, 179–184. [CrossRef]
12. Marshall, R.; Atlas, L.; Putner, T. The preparation and performance of thin film thermocouples. *J. Sci. Instrum.* **1966**, *43*, 144–149. [CrossRef]
13. Yang, L.; Zhao, Y.; Feng, C.; Zhou, H. The influence of size effect on sensitivity of Cu/CuNi thin- film thermocouple. *Phys. Procedia* **2011**, *22*, 95–100. [CrossRef]
14. Chen, Y.; Jiang, H.; Zhao, W.; Zhang, W.; Liu, X.; Jiang, S. Fabrication and calibration of Pt–10%Rh/Pt thin film thermocouples. *Measurement* **2014**, *48*, 248–251. [CrossRef]
15. Bhatt, H.D.; Vedula, R.; Desu, S.B.; Fralick, G.C. Thin film TiC/TaC thermocouples. *Thin Solid Films* **1999**, *342*, 214–220. [CrossRef]
16. Tian, X.; Kennedy, F.E.; Deacutis, J.J.; Henning, A.K. The development and use of thin film thermocouples for contact temperature measurement. *Tribol. Trans.* **1992**, *35*, 491–499. [CrossRef]

17. Shinozuka, J.; Obikawa, T. Development of cutting tools with built-in thin film thermocouples. *Key Eng. Mater.* **2004**, *257–258*, 547–552. [CrossRef]
18. Lüthje, H.; Bandorf, R.; Biehl, S.; Stint, B. Thin film sensor for wear detection of cutting tools. *Sensors Actuators A Phys.* **2004**, *116*, 133–136. [CrossRef]
19. Seibel, S. Vielfalt am laufenden Meter. *Kunststoffe* **2005**, *12*, 38–46.
20. Michaeli, W. *Extrusions-Werkzeuge für Kunststoffe und Kautschuk. Bauarten, Gestaltung und Berechnungsmöglichkeiten*; Hanser: München, Germany, 1991; ISBN 3446156372.
21. Tillmann, W.; Kokalj, D.; Stangier, D. Optimization of the deposition parameters of Ni-20Cr thin films on thermally sprayed Al_2O_3 for sensor application. *Surf. Coat. Technol.* **2018**, *344*, 223–232. [CrossRef]
22. Tillmann, W.; Kokalj, D.; Stangier, D.; Schöppner, V.; Benis, H.B.; Malatyali, H. Influence of Cr-Content on the thermoelectric and mechanical properties of NiCr thin film thermocouples synthesized on thermally sprayed Al_2O_3. *Thin Solid Films* **2018**, *663*, 148–158. [CrossRef]
23. Oliver, W.C.; Pharr, G.M. An improved technique for determining hardness and elastic modulus using load and displacement sensing indentation experiments. *J. Mater. Res.* **1992**, *7*, 1564–1583. [CrossRef]
24. DIN EN ISO 26443 German Institute for Standardization. *Fine Ceramics (Advanced Ceramics, Advanced Technical Ceramics)—Rockwell Indentation Test for Evaluation of Adhesion of Ceramic Coatings*; Beuth Verlag GmbH: Berlin, Germany, 2016.
25. Hochrein, T.; Alig, I. *Prozessmesstechnik in der Kunststoffaufbereitung*, 1st ed.; Vogel Business Media: Würzburg, Germany, 2011; ISBN 9783834331175.
26. Ait Aissa, K.; Achour, A.; Camus, J.; Le Brizoual, L.; Jouan, P.-Y.; Djouadi, M.-A. Comparison of the structural properties and residual stress of AlN films deposited by dc magnetron sputtering and high power impulse magnetron sputtering at different working pressures. *Thin Solid Films* **2014**, *550*, 264–267. [CrossRef]
27. Choudhary, R.K.; Mishra, S.C.; Mishra, P.; Limaye, P.K.; Singh, K. Mechanical and tribological properties of crystalline aluminum nitride coatings deposited on stainless steel by magnetron sputtering. *J. Nucl. Mater.* **2015**, *466*, 69–79. [CrossRef]
28. Bagcivan, N.; Bobzin, K.; Theiß, S. (Cr1−xAlx)N: A comparison of direct current, middle frequency pulsed and high power pulsed magnetron sputtering for injection molding components. *Thin Solid Films* **2013**, *528*, 180–186. [CrossRef]
29. Bobzin, K.; Nickel, R.; Bagcivan, N.; Manz, F.D. PVD-coatings in injection molding machines for processing optical polymers. *Plasma Process. Polym.* **2007**, *4*, S144–S149. [CrossRef]
30. Silva, F.J.G.; Martinho, R.P.; Baptista, A.P.M. Characterization of laboratory and industrial CrN/CrCN/diamond-like carbon coatings. *Thin Solid Films* **2014**, *550*, 278–284. [CrossRef]
31. Tillmann, W.; Hagen, L.; Hoffmann, F.; Dildrop, M.; Wibbeke, A.; Schöppner, V.; Resonnek, V.; Pohl, M.; Krumm, C.; Tiller, J.C.; et al. The detachment behavior of polycarbonate on thin films above the glass transition temperature. *Polym. Eng. Sci.* **2016**, *56*, 786–797. [CrossRef]
32. Goryacheva, I.G. *Contact Mechanics in Tribology*; Springer: Dordrecht, The Netherlands, 1998; ISBN 9789048151028.
33. Lugscheider, E.; Bobzin, K. The influence on surface free energy of PVD-coatings. *Surf. Coat. Technol.* **2001**, *142–144*, 755–760. [CrossRef]
34. Theiß, S. *Analyse Gepulster Hochleistungsplasmen zur Entwicklung Neuartiger PVD-Beschichtungen für die Kunststoffverarbeitung*; Shaker: Aachen, Germany, 2013; ISBN 9783844021127.
35. Vera-Sorroche, J.; Kelly, A.; Brown, E.; Coates, P.; Karnachi, N.; Harkin-Jones, E.; Li, K.; Deng, J. Thermal optimisation of polymer extrusion using in-process monitoring techniques. *Appl. Therm. Eng.* **2013**, *53*, 405–413. [CrossRef]

 *coatings*

Article

Influences of Oxygen Ion Beam on the Properties of Magnesium Fluoride Thin Film Deposited Using Electron Beam Evaporation Deposition

Gong Zhang [1], Xiuhua Fu [1,*], Shigeng Song [2,*], Kai Guo [3] and Jing Zhang [1]

[1] School of OptoElectronic Engineering, Changchun University of Science and Technology, Changchun 130000, China; zgoptics@126.com (G.Z.); zhangjing840225@cust.edu.cn (J.Z.)
[2] Institute of Thin Films, Sensors and Imaging, School of Engineering and Computing, University of the Westof Scotland, PA1 2BE Paisley, UK
[3] Optoelectronic component development, Beijing Institute of Control Engineering, Beijing 100094, China; kg_optics@126.com
* Correspondence: fuxiuhua@cust.edu.cn (X.F.); shigeng.song@uws.ac.uk (S.S.)

Received: 30 October 2019; Accepted: 3 December 2019; Published: 7 December 2019

Abstract: Magnesium fluoride (MgF_2) materials are commonly used for near/medium infrared anti-reflection optical coatings due to their low refractive index and wide-range transparency. Ion assistant deposition is an important technique to increase the density of MgF_2 and reduce absorption around 2.94 μm caused by high porosity and moisture adsorption. However, the excessive energy of Argon ion will induce a color center and; therefore, lead to UV/Visible absorption. In this paper, oxygen ion was introduced to reduce the color center effect in MgF_2 thin film deposited using electron beam evaporation with ion assistant. The films were deposited on Bk7 and single crystal silicon substrates under various oxygen ion beam currents. The microstructure, optical constant, film density, stress, and adhesion are investigated, including the absorption properties at the infrared hydroxyl (–OH) vibration peak. The results show that as the oxygen ion beam current increases, the absorption value at the position of the infrared OH vibration, defects, and stress of the film decrease, while the refractive index increases. However, MgF_2 has poor adhesion using oxygen ion-assisted deposition. Thin MgF_2 film without ion beam assistant was used as adhesive layer, high density, and low absorption MgF_2 film with good adhesion was obtained.

Keywords: MgF_2; color center absorption; density; crystal frequency; stress; adhesion

---

1. Introduction

Near/medium infrared detectors can work in both low light night vision and medium bands simultaneously, which can be widely used in military and civil fields [1–3]. The MgF_2 is a suitable material due to its low refractive index and transparency over the ultraviolet range to the far-infrared region [4,5]. However, the films easily absorb moisture in the atmosphere and relatively large tensile stress because of poor film structure, resulting in spectrum shift. The results show that increasing the deposition temperature of the substrate can increase the density of the film and reduce the instability caused by moisture absorption of MgF_2. When the deposition temperature of the substrate reaches 0.6 Tm [6] (where Tm is the melting temperature of the material), the aggregation density reaches a stable value of 0.9. Thereafter, the density will not increase with the increase of deposition temperature. Raising the temperature cannot completely solve the problem of the aggregation density [7,8]. Therefore, ion beam-assisted deposition is used to improve the density of the film. However, there are limitations to be considered in ion beam-assisted deposition. Both Matin Bischoff and Targove researches showed a technology which can produce high packing density and a low extinction coefficient by plasma(ion)-assisted deposition, but the pure fluorine introduced in the vacuum chamber will influence the environment and also reduce the life of the vacuum chamber. M. Kennedy research showed that xenon ion bombardment has a better result than Ar ion backfill oxygen at ultraviolet–visible wavelengths, but the infrared property of films has not yet been considered in this paper [9–11]. Dumasp's research showed that the use of Ar as an ion source to assist gas deposition increased the bulk density of MgF_2 films, but excessive particle energy led to a change in the stoichiometric ratio of Mg and F, resulting in F atom vacancies, which led to the absorption of ultraviolet and visible parts [12,13]. These studies further show that excessive Ar^+ ion energy can lead to more serious moisture absorption problems when MgF_2 thin films are evaporated by an electron gun [14].

The working band of near/mid infrared anti-reflective film contains the characteristic peak of water absorption within the band. Moisture absorption will greatly affect the performance of anti-reflective film. Therefore, it should be considered to increase the aggregation density of film while filling F^- ion vacancies. Sun discussed the effects of MgF_2 prepared at different temperatures on the spectral transmittance and absorption of the deep ultraviolet band produced by the MgO content in the film layer placed in air, which reduced the absorption of the color center [15]. It should be noted, the oxygen can reduce the F^- vacancy. In the paper the effect of oxygen ion source-assisted deposition on the properties of MgF_2 is studied. In order to obtain MgF_2 thin films with high density and low absorption by O^{2-} assisted deposition. The problem of poor bonding between MgF_2 thin films and Si substrates in the experimental process was also analyzed, and the ion source-assisted deposition technology was optimized to improve the adhesion of MgF_2 thin films on Si substrates assisted by O_2 ion source.

2. Experiment

2.1. Thin Film Preparation

The films were prepared by electron beam evaporation with O_2 ion beam-assisted deposition. The vacuum chamber was evacuated to a base pressure of less than 1.0×10^{-3} Pa. With a substrate temperature of 300 °C, thin films were deposited on crystal silicon (φ20 mm × 1 mm) and float glass (φ25 mm × 1 mm) with a deposition rate of 0.8nm/s with different ion beams. The substrates were cleaned by ultrasonication. The experiment used a quartz crystal to monitor the physical thickness of the film. The monitor was SQC-310 produced by Inficon Co., Ltd (Inficon, Shanghai, China). To investigate oxygen ion effects on MgF_2 film properties, samples were deposited under various oxygen ion currents (Kaufman Ion beam assistant) of 0, 50, and 80 mA at deposition temperature of 300 °C and 0.8 nm/s deposition rate. Sample thicknesses were controlled at 700 nm.

2.2. Film Characterization

The properties of MgF_2 samples deposited under various oxygen ion currents, such as the crystal frequency, optical transmittance, chemical phase of the film, surface roughness, and adhesion, were analyzed using the techniques listed in Table 1.

Table 1. Measurement process.

Group	Equipment and Accuracy	Purpose	Process
1	Inficon golden crystaluency ($\Delta f < 0.004657$ Hz)	Peaking density	1. Recorded the crystal frequency after coating at pressure of 1.0×10^{-3} Pa, recorded the crystal frequency, vented and recorded again. 2. The crystal was taken out and soaked in the ionized water for 36 h to make the crystal oscillator fully absorb water and then the frequency of the crystal oscillator was displayed in the XTC-3100 quartz monitoring system until the frequency stabilized. 3. Thereafter, the crystal was placed in the chamber in which it was pumped to 1.0×10^{-3} Pa again at a baking temperature of 300 °C, and the vibration frequency of the crystal piece was recorded.
2	SHIMADZU UV-3150 ($\Delta T < 0.0002\%$ T)	UV–Visible spectrum	After deposition, the film deposited on BK7 substrate was measured, optichar software (version 9.51) was used to fit the transmittance curve and calculate the optical constant. The Cauchy dispersion module homogeneous layer with a smooth surface was used to fit the optical constant.
3	Agilent VR610 ($\Delta T < 0.05\%$ T)	Mid-infrared spectrum	After deposition, the film transmittance deposited on Si substrate was measured, and the –OH absorption of different samples were analyzed.
4	Bruker D8 Advance XRD ($\Delta\theta < \pm 0.001°$)	Chemistry phase of film	X-ray diffraction studies were performed with a Bruker D8 Advance XRD instrument equipped with Cu Ka source, the scan type was "absolute" and scan mode was continuous
5	Zygo interferometer ($\Delta\lambda < 1/2000\lambda$)	The films' stress	The Zygo interferometer tested the Δpower value then calculated the stress of film.
6	Adhesive experiments	3M adhesive tape	Test was done using 3M scotch 610 test tape: Stickiness of (10 ± 1) N/25mm, and keeping tape pull angle at 90° to the film surface.

2.3. Some Background for Characterizations

2.3.1. Quartz Frequency Measurement of Density

The refractive index of the film will change with the absorption of moisture by the pores within the film. Usually, the packing density or porosity of thin films can be calculated using linear interpolation formula according to the change of refractive index [16]. The refractive index of dense MgF_2 film is about 1.38, which is very close to the refractive index of water 1.33. The refractive index change of the film layer after moisture absorption is very small, and it is difficult to calculate the density of the film through the change of refractive index. Therefore, in this study, the density of porous MgF_2 was calculated based on the measurement of the mass change of film before and after moisture absorption. A quartz crystal plate is very sensitive to the change of mass. Therefore, the density or porosity of the film layer can be obtained accurately through the change of frequency of quartz crystal plate before and after the absorption of water. The derivation process of the specific calculation is as follows [17].

Its natural frequency is inversely proportional to the thickness t and is proportional to the frequency constant N. The relationship is shown in Equation (1)

$$f = N/t \tag{1}$$

$$\Delta f = -\frac{N\Delta t}{t^2} \tag{2}$$

Differentiating Equation (1) with respect to the thickness to get the crystal frequency:

After MgF_2 film is deposited, its density is close to that of quartz and the thickness of the coated film during the experiment is much smaller than that of the quartz crystal oscillator, the increase in film thickness can be approximated as the change of the thickness of the quartz crystal.

$$\Delta m = A \bullet \rho_M \bullet \Delta t_M = A \bullet \rho_Q \bullet \Delta t \tag{3}$$

where Δm represents the mass change, A is the crystal coating area, and ρ_M and ρ_Q are the film density and the quartz density, respectively. By substituting into Equation (2), the relationship between the change in the quartz crystal frequency and the film thickness can be obtained by Equation (4).

$$\Delta f = -\frac{\rho_M}{\rho_Q} \bullet \frac{f^2}{N} \Delta t_M \tag{4}$$

It can be seen from the formula that the decrease in the quartz crystal frequency is proportional to the thickness and density of the coated film.

Assume that the initial frequency of the crystal is f_1, the frequency after water absorption is f_1^*, the bulk density of the film is ρ_s, and the density is shown in Equation (5),

$$p = \frac{\Delta f_1}{\Delta f_1 + \rho_s \Delta f_1^*} \tag{5}$$

According to Equation (5), the density under various conditions can be easily calculated.

2.3.2. Zygo Interferometry for Measuring Film Stress

Stoney proposed calculating the stress of a thin film by measuring the radius of the deformation curvature of the film and the substrate surface. When the actual thickness of the film is considered to be infinitesimal relative to the substrate, both the film and the substrate can be considered as homogeneous and isotropic materials [18]. The Stoney formulate is shown below:

$$\sigma_f = \left(\frac{E_s}{1 - v_s}\right)\frac{t_s^2}{6Rt_f} \tag{6}$$

where, Es and V_s are the elastic modulus and Poisson's ratio of the substrate, respectively. T_s and t_f are the thickness of the substrate and the film, respectively. R is the radius of curvature of the substrate. Based on the interference principle, the Zygo interferometer calculates the curvature change of the measured object by comparing the interference fringe generated by the optical path difference between the measured light and the reference light reflected from the measurement plane [19]. By processing the test results of the Zygo interferometer, the power value indicating the apparent degree of surface deformation can be obtained. The power is showed in Equation (7):

$$power = \frac{D_s{}^2}{8R} \tag{7}$$

where D_s is the diameter of the substrate, and the radius of curvature before and after coating can be expressed by Equation (8),

$$\frac{1}{R_2} - \frac{1}{R_1} = \frac{8}{D_s^2}(power_2 - power_1) \tag{8}$$

Combining Equation (6), the surface stress of the thin film can be expressed by Equation (9),

$$\sigma = \frac{4Es}{3(1-v_s)}\frac{t_s^2}{t_f D_s{}^2}\Delta power \tag{9}$$

3. Result and Analysis

3.1. Packing Density Character

The variation of crystal oscillator frequency with time after coating is shown in the Figure 1.

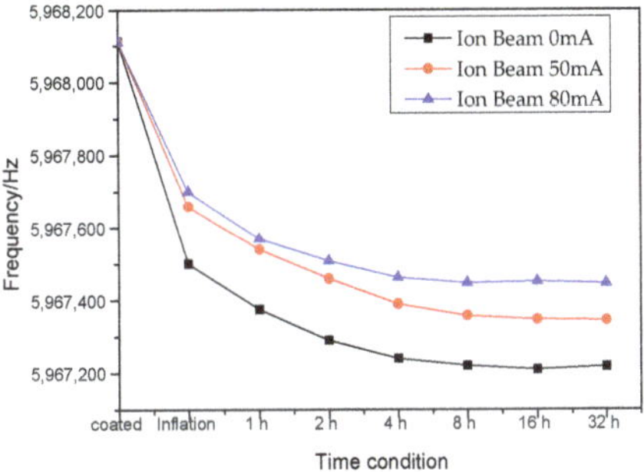

Figure 1. Crystal frequency changed with time in different ion beams (the black, red and blue lines represent the frequency with ion beam assistant condition in ion beams 0, 50, and 80 mA, respectively).

As can be seen from the figure, during the venting of the vacuum chamber to the atmospheric state, the frequency of the crystal vibrator is significantly decreased. As the immersion time in the deionized water increases, the frequency of the crystal vibrator decreases continuously, and finally reaches a stable value. The value in the crystal vibrator frequency decreases as the ion beam density increases. According to the formula listed in Section 2.1, the crystal oscillator frequency variation value is used to calculate the packing density of the film. The aggregation densities under different processing conditions are shown in Table 2.

It can be seen from the table that as the oxygen ion beam increases, the density of the film increases. Therefore the oxygen ion-assisted deposition can be an effective technique to reduce the porosity of MgF_2 film.

As discussed above, the deposited MgF_2 films on quartz crystal were immersed in DI water for a certain time to allow the void of film to be filled by water, and oscillation frequencies of the crystal were recorded. Then the samples were placed back in a chamber under high vacuum and at high temperature to remove the water trapped in the void. The oscillation frequencies of crystal were also recorded. Table 3 below shows the frequencies of crystal with MgF_2 film with and without adsorbed moisture, and the frequency differences.

Table 2. The density of the MgF_2 films produced at various oxygen ion currents.

Ion Beam	Density
0	0.89
50	0.905
80	0.92

Table 3. The crystal frequency under various conditions.

Oxygen Ion Current	0 mA	50 mA	80 mA
Before water immersion (A)	5,968,113.2	5,968,115.3	5,968,110.7
After water immersion (B)	5,967,217.3	5,967,345.3	5,967,447.3
Under vacuum at 300 °C (C)	5,967,982.4	5,968,003.2	5,968,092.6
Frequency differences (C–B)	761.5	657.9	645.3
Frequency differences (C–A)	−130.8	−112.1	−18.1

As the ion beam current increases, the frequency difference between the crystal oscillators before and after water immersion decreases (C–A in Table 4). When the ion beam density is 80 mA, the frequency of the crystal oscillator before (A) and under vacuum after the immersion (C) is very close. It shows that under ion beam condition of 80 mA, the main reason for the change of the crystal vibration frequency is that water vapor penetrates into the pores of the film, and only physical adsorption occurs mainly. Water molecules trapped in the pores with strong bonding are almost negligible. When oxygen ion current is 0 mA, the difference between the crystal oscillators frequency under vacuum at 300 °C (C in Table 4), and the frequency just after the completion of MgF_2, is much higher compared to the one under 80 mA oxygen ion current, which means there is a much stronger bonding force between water molecule and MgF_2. This indicates that the water vapor is not only physically adsorbed within the deposited film on the crystal oscillator, but also undergoes a chemical reaction to form a strong bond. Through the above tests, it was shown that oxygen ion bombardment introduces modifications of MgF_2 film when the MgF_2 film was deposited using electron beam evaporation under oxygen ion assistant. The participation of oxygen ion would supplement the crystal defects, improve the crystal structure, and cause the modification of the MgF_2 film surface state, including the surface of pores in film (e.g., Mg suspending bond). Oxygen ion also improves the density of the film. From a macroscopic point of view, the refractive index of the film increases, absorption and scattering decrease, and tensile stress decreases.

Table 4. Transmittance change.

Film Condition	Ion Beam 0 mA	Ion Beam 50 mA	Ion Beam 80 mA
After coated	3.22%	1.74%	0.83%
32 h dipped	5.65%	2.64%	1.52%
After vacuum baked	5.2%	1.98%	0.96%

3.2. Optical Properties

Visible/near infrared and mid-infrared transmittance were measured. Visible/near infrared transmittance was used to fit the optical constant and the mid-infrared transmittance was used to analyze the absorption of hydroxyl near 2940 nm.

3.2.1. The Visible/Near Infrared Analysis

The Visible-near infrared band film transmittance curve is shown in the Figure 2.

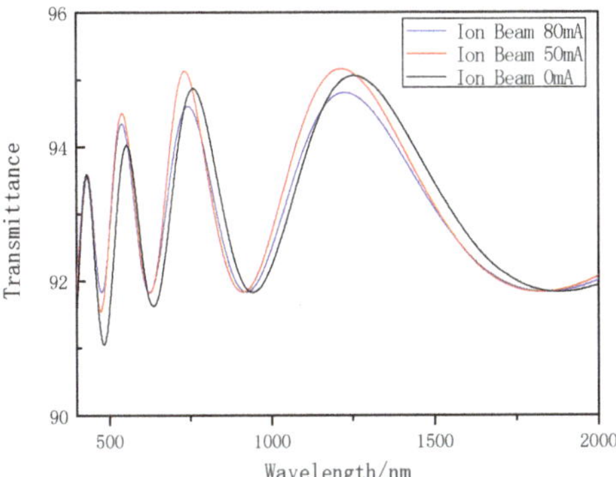

Figure 2. The transmittance of different ion beams in the Vis-infrared band (the black, red, blue lines represent transmittance with the ion beam assistant condition in ion beams 0, 50, and 80 mA, respectively).

It can be seen from the figure that there are significant differences in the refractive index and absorption of MgF_2 films under different ion beam flow conditions. The optical constant dispersion curve of the film obtained by the full spectrum fitting method is shown in Figure 3.

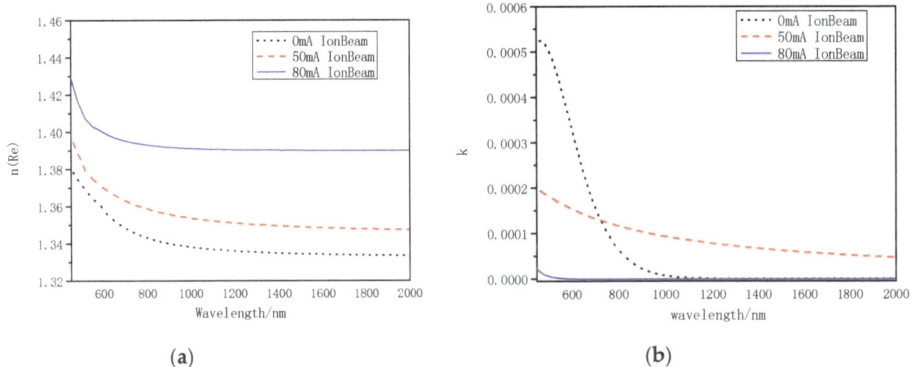

(a)   (b)

Figure 3. Optical constant dispersion curve (**a**) is the refractive index, (**b**) is the extinction coefficient the black, red and blue lines represent optical constant with the ion beam assistant condition in ion beams 0, 50, and 80 mA, respectively.

It can be seen that the O_2 ion beam causes the refractive index to increase. It was because of the presence of MgO compound in MgF_2 film. Using the Wiener bounds model, the amount of MgO roughness was estimated as 13.1%.

3.2.2. Hydroxyl Vibration Absorption

The infrared transmittance curve of the single crystal Si substrate [20] and the transmittance of the MgF_2 film prepared under different ion source conditions are shown in Figure 4.

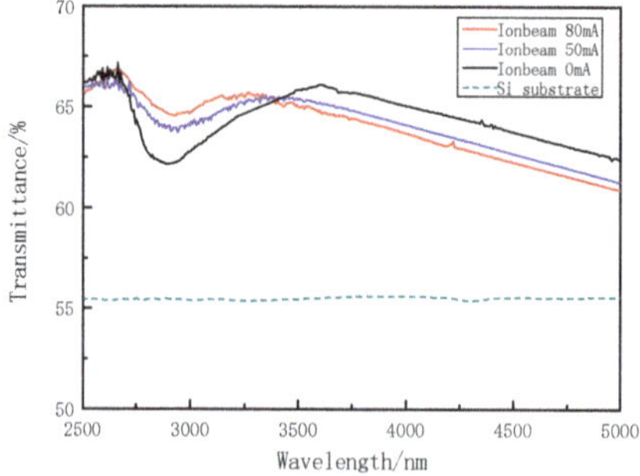

Figure 4. The transmittance of MgF$_2$ and substrate (the black, red, blue lines represent mid-infrared transmittance with the ion beam assistant condition in ion beams 0, 50, 80 mA, respectively. The dashed line is Si substrate).

It can be seen that the spectral transmittance of the monocrystalline Si substrate at the hydroxyl vibration position of 2770~3200 nm also fluctuates, indicating that the compounds containing –OH exist on the surface of the Si substrate [21]. The spectral transmittance of MgF$_2$ prepared under different ion source conditions at the band of 2770~3200 nm is significantly decreased, and the characteristic absorption peak of –OH appears at 2940 nm. When the substrate was immersed in deionized water, the spectral transmittance decreases in accordance with the frequency change of the crystal oscillator of Section 3.1. As the immersion time increases, the spectral transmittance decreases continuously and is substantially stable after 32 h. After baking under vacuum for two hours, the infrared spectrum of 2500~5000 nm was retested. The average reduction values of the spectral transmittance before and after baking in the range of 2770~3200 nm are shown in Table 4.

From the table it can be seen that ion beam-assisted deposition can effectively reduce the absorption of MgF$_2$ in the range of 2770 to 3200 nm, and the spectral transmittance decreases the least when the ion beam is 80 mA. After baking under vacuum, the decrease of the spectral transmittance of MgF$_2$ with an ion beam density of 80 mA at 2770 to 3200 nm is basically the same as that before immersion, indicating that the moisture physically adsorbed in the film causes the decrease in the spectral transmittance of the film. While, without using ion bombardment, the spectral transmittance of the film is only changed by 0.4% after baking under vacuum conditions. After baking in vacuum, the film still contains compounds with hydroxyl groups.

This indicates that the decrease of absorption of MgF$_2$ thin film prepared by oxygen ion source-assisted deposition near 2940 nm is not only due to the increase of concentration density of the film layer, but also due to the recombination of oxygen ion and F$^-$ vacancy caused by Mg suspension bond during deposition, preventing the suspension bond from forming other compounds with hydroxyl in water vapor.

3.3. XRD

According to the result of spectrum test, the oxygen ion and Mg suspending bond fill the film defects, reduce the colour center absorption, and reduce the hydroxyl absorption of film by filling the F$^-$ vacancy with O particles. To analyze the phase change of film, an XRD measurement was taken, the result are shown in Figure 5.

The XRD results show that MgF$_2$ film exhibits a polycrystalline state, and the film has a remarkable crystal orientation, with the maximum intensity of the diffraction peak in the <110> direction. It can be seen that the location of the diffraction peak without assisted deposition by oxygen ion source is consistent with that of the standard PDF (powder diffraction file) card, as the oxygen ion beam flow increases, the diffraction peak positions in each direction shifts to the right, the intensity of the diffraction peak decreases, and the diffraction peak gradually widens as the ion beam flow increases.

When the ion beam was increased to 80 mA, the MgF_2 diffraction peak in the <220> direction and significantly broadened. The diffraction peak became inconspicuous. In addition, the characteristic diffraction peak of MgO appears at positions where the diffraction angle (2θ) values are 36.703° and 62.219°, indicating that under the beam condition the O ion compounds with Mg in the film to form a MgO crystal.

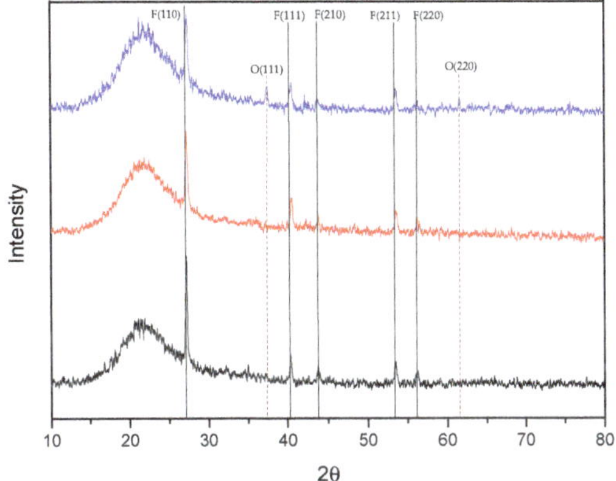

Figure 5. XRD measurement of different ion beams (the blue, red, and black represent ion beams 80, 50, and 0 mA, respectively; the vertical lines represent the theoretical positions of the peaks for MgF2 and MgO; F represents MgF_2, O represents MgO).

The comparison parameters of the crystal plane spacing of MgF_2 thin film and PDF standard card are as per Table 5.

Table 5. The crystal plane spacing of sample and PDF standard value.

Spacing (d) <hkl>	Literature ICCD41-1443	Sample 1 Ion Beam 0 mA	Sample 2 Ion Beam 50 mA	Sample 3 Ion Beam 80 mA
<110>	3.2670	3.2668	3.2683	3.2686
<111>	2.2309	2.2299	2.2312	2.2315
<210>	2.0672	2.0670	2.0673	2.0674
<211>	1.7112	1.7113	1.7120	1.7124
<220>	1.6335	1.6338	1.6343	1.6345

The lattice constants of the MgF_2 film and the PDF reference card prepared under different process conditions were calculated by Jade software and are shown in Table 6.

Table 6. The crystal character calculates by MDI Jade.

Lattice	Literature ICCD41-1443	Sample 1 Ion Beam 0 mA	Sample 2 Ion Beam 50 mA	Sample 3 Ion Beam 80 mA
a (nm)	0.46200	0.46220	0.4295	0.4635
C (nm)	0.30509	0.30622	0.3041	0.3032

As can be seen, with the increase of oxygen ion beam, the crystal axis cell becomes smaller and the spacing between the crystal planes decreases. The oxygen ion source-assisted deposition indicates that the crystal cells gather more closely during the growth of the film.

3.4. Adhesion

The tape peeling method was used for adhesion test, where tape is attached to the sample and then peeled off. The adhesion is then judged on the extent of the film delamination. The measurement process is simple and has good repeatability. The test was done using 3M scotch 610 test tape: Stickiness of (10 ± 1) N/25mm, and keeping tape pull angle at 90° to the film surface [22].

The film without O_2 ion assistant deposition can endure being peeled off 20 times. When the O_2 ion beam was 50 mA, the film delaminated after being peeled off 15 times. When the ion beam increased to 80 mA the film delaminates after being peeled off 12 times. Indicating the film adhesion decreases with increasing ion beam. The adhesion of the film is mainly affected by stress and inter-material adsorption [23]. Thus, the stress was discussed as follows.

The changes in curvature of the MgF_2 film before and after deposition obtained by a Zygo interferometer are shown in Table 7. According to the formula in Section 2.2, the stress of the film is 986, 436, and 357 MPa, respectively. It can be seen that the tensile stress of the film decreases as the ion beam density increases.

Table 7. Curvature of substrate before and after coating.

Ion Current	Power before Coated	Power after Coated
Ion Beam 0 mA	0.091	0.866
Ion Beam 50 mA	0.091	0.425
Ion Beam 80 mA	0.091	0.365

The stress test results show that when the ion beam is 80 mA, the tensile stress of the film is at the minimum, and the stress of the film should not be the main cause for poor adhesion. It can be seen from the infrared spectrum test results that at 2770~3200 nm the Si substrate has obvious absorption, indicating that the surface of the Si substrate contained the OH root compound. When the MgF_2 film was modified without using oxygen ions, there were obvious F^- vacancies in the film layer. The free Mg^{2+} ions were combined with the hydroxyl of the Si substrate. The film and the substrate were combined by the bonding force. The MgF_2 film substrate, in which the F ionic vacancies were filled by the O particles, was combined by van der Waals force. Therefore, the film–substrate adhesion is much smaller than the bonding force [24,25]. In order to obtain a MgF_2 film with good adhesion to the substrate, and effectively reducing the absorption of water molecules, ion beam-assisted deposition was not used at the beginning of up to 50 nm thickness of deposition, and a film (transition layer film) having F^- vacancies was obtained. The –OH compounds formed by bonding with the Si substrate made the film adhere well to the substrate. Next, O ion bombardment by increasing the beam density to 80 mA was used to fill the F^- ion vacancies in the film during deposition. The comparison of 3M adhesive experiments before and after adjusting the ion source is shown in Figure 6.

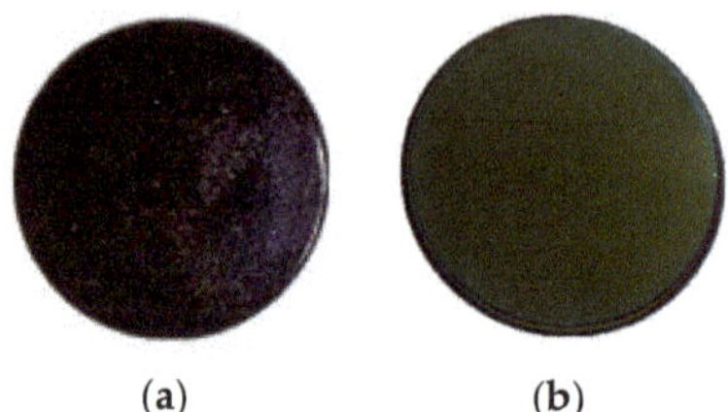

(a)   (b)

Figure 6. Sample condition after tape pull test: (**a**) Without transition layer which was produced by electron beam vapor deposition and O_2 IAD; (**b**) the first transition 50 nm deposited using only electron beam vapor deposition, then deposition completed with IAD.

In addition, the comparison of transmittance before and after adjusting the ion source is shown in Figure 7.

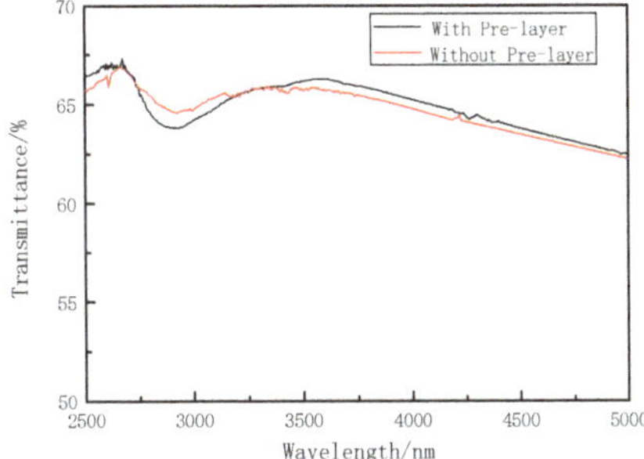

Figure 7. The transmittance of MgF_2 (the black and red line represent mid-infrared transmittance with 50 nm pre-layer and without pre-layer, respectively.)

4. Conclusions

The influence of oxygen ion source on MgF_2 film's visible near-infrared band and mid-wave infrared absorption was studied. As the oxygen ion beam flow increases, the film aggregation density increases. Due to the oxygen ions filling into the colour center defects, generated by F^- vacancies, an MgO compound is formed in the film, so that the absorption value of the film in the visible range is reduced, and the refractive index is increased. In the infrared light portion, as the oxygen ion beam increases, the F^- ionic vacancies are filled with O, preventing the Mg^{2+} ions from recombining with the –OH in the air, and reducing the absorption of the film in the infrared. However, the stress and adhesion test results show that as the oxygen ion beam flow increases, the bonding force between the film and the substrate changes from chemical bonding force to van der Waals force, and the tensile stress of the MgF_2 film decreases, which lead to a bad adhesion for MgF_2 film on Si substrate. A solution of two steps of deposition was proposed to solve the problem of poor adhesion: At the early stage of deposition of the MgF_2 film, a very thin adhesion layer without ion beam-assisted was formed and allowed further MgF_2 film to bond to the substrate with good adhesion, then the O^+ ion current for further MgF_2 deposition was increased to allow O to fill the F anion vacancies efficiently, and reduce the absorption. The experimental results demonstrate that MgF_2 film with low absorption, high stability, and good adhesion was achieved. The effect of the thickness of the thin adhesive MgF_2 layer deposited without ion beam assisted requires further investigation.

Author Contributions: Conceptualization, G.Z. and X.F.; methodology, G.Z.; software, G.Z.; validation, S.S., K.G. and G.Z.; formal analysis, S.S.; investigation, J.Z.; resources, J.Z.; data curation, S.S.; writing—original draft preparation, G.Z.; writing—review and editing, S.S.; visualization, K.G.; supervision, X.F.; project administration, X.F.

Funding: This research received no external funding.

Conflicts of Interest: The authors declare no conflicts of interest.

References

1. Pham, T.; Du, W.; Tran, H.; Margetis, J.; Tolle, J.; Sun, G.; Soref, R.A.; Naseem, H.A.; Li, B.; Yu, S.-Q. Systematic study of Si-based GeSn photodiodes with 26 µm detector cutoff for short-wave infrared detection. *Opt. Express* **2016**, *47*, 1–18. [CrossRef] [PubMed]
2. Zhang, A.; Kim, H.; Cheng, J.; Lo, Y.-H. Ultrahigh responsivity visible and infrared detection using silicon nanowire photo transistion. *Nano Lett.* **2010**, *10*, 2117. [CrossRef] [PubMed]
3. Zhang, T.; Lin, C.; Chen, H.; Sun, C.; Lin, J.; Wang, X. MTF measurement and analysis of linear array HgCdTe infrared detectors. *Infrared Phys. Technol.* **2018**, *8*, 123–127. [CrossRef]
4. Dodge, M.J. Refractive properties of magnesium fluoride. *Appl. Opt.* **1984**, *23*, 1980–1985. [CrossRef] [PubMed]

5.	Williams, M.W. Optical properties of magnesium fluoride in the vacuum ultraviolet. *J. Appl. Phys.* **1967**, *38*, 1701. [CrossRef]

6.	Al-Douri, A.A.J.; Heavens, O.S. The role of rate of deposition and substrate temperature on the structure and loss of zinc sulphide and magnesium fluoride films. *J. Phys. D* **1983**, *16*, 927–935. [CrossRef]

7.	Zhang, Y. Research on Preparation of low absorption infrared coatings and its environmental stability and reliability. Ph.D. Thesis, Institute of Optics and Electronics, Chinese Academy of Sciences, Beijing, China, June 2019. (In Chinese).

8.	Bischoff, M.; Stenzel, O.; Friedrich, K. Plasma-assisted deposition of metal fluoride coatings and modeling the extinction coefficient of as-deposited single layers. *Appl. Opt.* **2011**, *50*, 232–238. [CrossRef]

9.	Guenther, K.H.; Pulker, H.K. Electron microscopic investigations of cross sections of optical thin films. *Appl. Opt.* **1976**, *15*, 2992–2997. [CrossRef]

10.	Targove, J.D. The ion-assisted deposition of optical thin films. Ph.D. Thesis, The University of Arizona, Arizona, Tucson, AZ, USA, 1987.

11.	Kennedy, M.; Ristau, D.; Niederwald, H.S. Ion beam-assisted deposition of MgF_2 and YbF_3 films. *Thin Solid Film.* **1998**, *333*, 1–2. [CrossRef]

12.	Dumas, L.; Quesnel, E. Characterization of magnesium fluoride thin films produced by argon ion beam-assisted deposition. *Thin Solid Film.* **2001**, *382*, 61–68. [CrossRef]

13.	Wilbrandt, S.; Stenzel, O. Combined in situ and ex situ optical data analysis of magnesium fluoride coatings deposited by plasma ion-assisted deposition. *Appl. Opt.* **2011**, *50*, 5–10. [CrossRef] [PubMed]

14.	Booster, J.L.; Voncken, J.H.L. Characterization of hydroxyl-bearing magnesium fluoride containing physically bound water. *Powder Diffr.* **2002**, *17*, 112–118. [CrossRef]

15.	Sun, J.; Shao, J. Effects of substrate temperatures on the characterization of magnesium fluoride thin films in deep-ultraviolet region. *Appl. Opt.* **2014**, *53*, 1298–1305. [CrossRef] [PubMed]

16.	Ogura, S.; Sugawara, N.; Hiraga, R. Refractive index and packing density for MgF_2 films: Correlation of temperature dependence with water sorption. *Thin Solid Film.* **1975**, *3*, 3–10. [CrossRef]

17.	Forman, P.F. *The Zygo Interferometer System*; Hopkins, G.W., Ed.; SPIE: Bellingham, WA, USA, 1979; Volume 192.

18.	Wang, J.; Shrotriya, P.; Kim, K.S. Surface residual stress measurement using curvature interferometry. *Exp. Mech.* **2006**, *46*, 39–46. [CrossRef]

19.	Ottermann, C.; Otto, J.; Jeschkowski, U. Stress of TiO_2 thin films produced by different deposition techniques. *MRS Proc.* **1993**, *308*, 69. [CrossRef]

20.	Daniel, C. Durable 3–5 micron transmitting infrared window materials. *Infrared Phys. Technol.* **1998**, *1998 39*, 185–201.

21.	Coker, D.; Reimers, J.; Watts, R. The infrared absorption spectrum of water. *Aust. J. Phys.* **1982**, *35*, 62. [CrossRef]

22.	Lu, N.H. Adhesion test for dry film photoresist on copper by the peeling method. *Circuit World* **1989**, *15*, 47–49. [CrossRef]

23.	Chapman, B.N. Thin-film adhesion. *J. Vac. Sci. Technol.* **1974**, *11*, 106–113. [CrossRef]

24.	Bikerman, J.J. Cause of poor adhesion boundary: Weak boundary layers. *Ind. Eng. Chem.* **1967**, *59*, 40–44. [CrossRef]

25.	Bikerman, J.J. The fundamentals of tackiness and adhesion. *J. Colloid Sci.* **1947**, *2*, 163–175. [CrossRef]

Article

Polarization Controlled Dual Functional Reflective Planar Metalens in Near Infrared Regime

Yuhui Zhang [1], Bowei Yang [1], Zhiying Liu [1,2,3] and Yuegang Fu [1,2,3,*]

[1] School of Optoelectric Engineering, Changchun University of Science and Technology, Changchun 130022, China; 2018200041@mails.cust.edu.cn (Y.Z.); 2019200168@mails.cust.edu.cn (B.Y.); lzy@cust.edu.cn (Z.L.)
[2] Key Laboratory of Optoelectronic Measurement and Optical Information Transmission Technology of Ministry of Education, Changchun University of Science and Tchnology, Changchun 130022, China
[3] Key Laboratory of Advanced Optical System, Design and Manufacturing Technology of the Universities of Jilin Province, Changchun 130022, China
* Correspondence: fuyg@cust.edu.cn

Received: 12 March 2020; Accepted: 11 April 2020; Published: 15 April 2020

Abstract: The metalens has been a hotspot in scientific communications in recent years. The polarization-controlled functional metalens is appealing in metalens investigation. We propose a metalens with dual functions that are controlled by polarization states. In the first design, when applied with x- and y-polarized light, two focal spots with different focal lengths are acquired, respectively. The proposed metalens performs well when illuminated with adjacent wavelengths. In the second design, the reflected light is focused when applied with x-polarized light, and when applied with y-polarized light, the reflected light is split into two oblique paths. We believe that the results will provide a new method in light manipulation.

Keywords: polarization controlling; dual functional-metalens; focusing; splitting

1. Introduction

Metamaterials are novel artificial structures that are designed for specified functions like negative refractive index [1–3], polarization conversion [4–6] and perfect absorption [7–9]. Metasurfaces possess the advantages of small losses and are easy to manufacture [10,11], and have been attractive in recent years. Among various metasurfaces, metalens with its unit cell interacting with incident light and introducing abrupt phase shift along the interface of the metasurface have been widely presented to achieve the function of focusing [12,13] and anomalous reflection [14–16]. Lots of models like the nanoslit [17], nanohole [18,19], and graphene ribbons [20–24] have been introduced in metalens design, where 2π phase shift resulting from the designed antennas is needed in controlling the wavefront. Although a polarization-independent metalens [12,25] and polarization-conversion metalens [26] have been reported in former works, and the designed metalenses are always adjusted to manipulate different kinds of incident waves with various structures [27–30], the metalens, with its function controlled by incident polarization state [31,32], has not been fully investigated.

In this paper, a metal-insulator-metal (MIM) structure is proposed to achieve the polarization controlled dual functional two-dimentional (2D) cylindrical metalens, which consists of rectangle gold antennas and a gold mirror spaced by a dielectric layer. The proposed metalens possesses two functions that can be realized when illuminated with x- and y-polarized light. The x-direction and y-direction length of the rectangle gold antenna are adjusted to control the phase and reflectance of the reflected light, because the length of the dipole resonances along x- and y-directions are determined by the x-direction and y-direction lengths, respectively. We first investigate the influence of the y-direction length or x-direction length on the dipole along the x or y-direction. By presenting the phase and reflectance of the reflected light for x-polarization incidence with different x-direction and y-direction lengths of the rectangle antenna, we find that the y-direction length of the rectangle has little influence on the phase and reflectance in most of its length range. Then we apply the phase approach to design the polarization-controlled metalens, in which the y-direction or x-direction length is fixed as 100 nm when varying the x-direction or y-direction length. We design a metalens with the focal length of $F = 5$ μm and $F = 5$ μm when applied with x- and y-polarized light, which focus well as desired. Then we validate the approach by presenting two metalenses with focal length of 5 and 15 μm for x- and y-polarization incidence only. The y-direction and x-direction lengths are fixed as 100 nm. The focusing effects agree well with that of the proposed polarization-controlled metalens. The proposed metalens works well within a broadband of wavelengths that range from 750 to 850 nm. We also design the metalens to split y-polarized normal incident light and focus the x-polarized light. Therefore, the dual functional metalens can be achieved by tuning the incident polarization state.

2. Design and Simulation Method

Figure 1 illustrates the schematic of the proposed 2D cylindrical lens, which is a metal-insulator-metal structure to form a Fabry–Perot cavity to enhance the interaction between light and resonance antenna. The resonance antenna and the bottom mirror are selected as Au with data acquired from [33]. The dielectric spacer is chosen as MgF$_2$ with a refractive index of 1.892 [14]. The thicknesses of Au antenna, MgF$_2$ spacer and Au mirror shown in Figure 1b are set as $t_a = 30$ nm, $t_d = 50$ nm and $t_s = 130$ nm, respectively. The top view of the unit cell of the metalens is shown in Figure 1c. The period in x- and y-directions are $p_x = p_y = 200$ nm. The resonance antenna is a rectangle, which possesses dipole resonances along both the x- and y-directions. The rectangle lengths along x- and y-directions are variable to control the phase and reflectance of the reflected light for x- and y-polarization incidences, respectively. Thus, we can separately control the wavefront of the reflected x- and y-polarized light by configuring the rectangle lengths along x- and y-directions. The working mechanism of the proposed metalens is shown in Figure 1a, where different polarization state leads to different function. For numerical analysis, all simulations are carried out by using the finite-difference time-domain software (Lumerical FDTD solutions 8.15.736.0). PML boundary condition is applied in z- and x- directions. Periodic boundary condition is applied in the y-direction. The minimal mesh size is 4 nm.

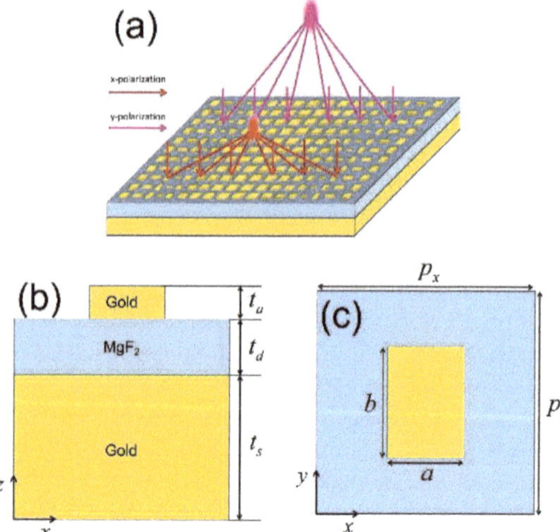

Figure 1. Schematic of the proposed metalens. (**a**) The fragment, (**b**) cross section view and (**c**) top view of the proposed metalens.

Generally, the influence of the y-direction length *b* or x-direction length *a* on the dipole along the x or y-direction cannot be ignored, which results in amplitude and phase deviations for the x- or y-polarization incidences. To explore this, we choose the working wavelength as 800 nm. We show the phase and reflectance of the reflected light for x-polarization incidence in Figure 2 with different lengths *b* and *a* of the rectangle antenna. A near 2π phase shift can be acquired when the length *a* increases from 10 to 190 nm, and the length *b* has little influence on it in most of the length range as shown in Figure 2a,b. Due to the symmetry property of the structure, an identical performance can be acquired for y-polarization incidence.

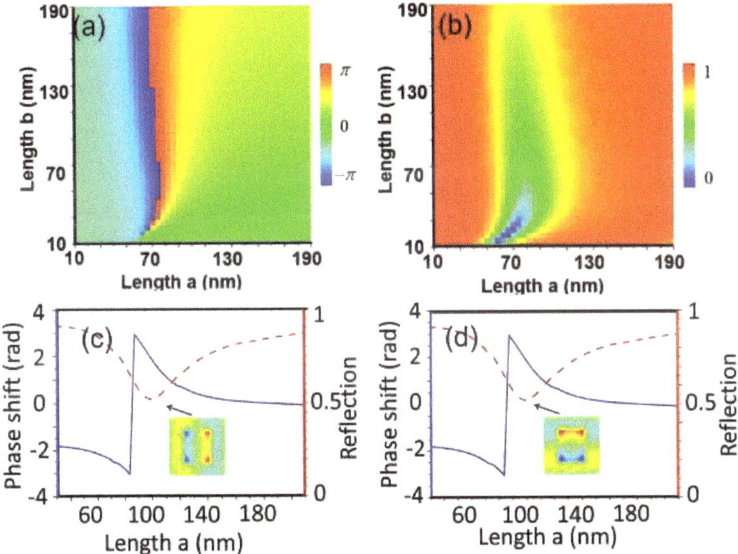

Figure 2. (**a**) Phase and (**b**) reflectance of the reflected light for x-polarization incidence with different lengths *b* and *a*. Phase shift and reflectance of reflected wave for (**c**) x- and (**d**) y-polarization incidences. Insets are z-component electric field distributions corresponding to (**c**) horizontal dipole and (**d**) vertical dipole.

Figure 2c shows the phase and reflectance of the reflected light for x-polarization incidence when x-direction length *a* increases from 10 to 190 nm and the y-direction length *b* is fixed as 100 nm. Figure 2d shows the phase and reflectance of the reflected light for y-polarization incidence when y-direction length *b* increases from 10 to 190 nm and the x-direction length *a* is fixed as 100 nm. The insets show the z component electric field distributions at the resonant length. Two dipole resonances along x- and y-directions are acquired, together with the Fabry–Perot resonance, a near 2π phase shift can be achieved. We use this approach to design the metalens that possesses different functions controlled by the polarization states of the incident light.

3. Results and Discussion

As illustrated above, we utilize the phase shift in Figure 2c,d to design the focusing metalens, which possesses the focal length of $F = 5$ μm for x-polarization incidence and $F = 15$ μm for y-polarization incidence. To design a metalens to focus incident light, the phase profile of the metalens should follow the expression [34]:

$$\varphi(x) = \frac{2\pi}{\lambda_0}\left(\sqrt{x^2 + F^2} - F\right) \tag{1}$$

where x is the horizontal position from the center of the metalens, Δx is the horizontal shift of the focal point, λ_0 is the incident wavelength, and F is the focal length. Based on Equation (1), we calculate the phase distribution curves for $F = 5$ μm and $F = 15$ μm, and show them in Figure 3a,b. The corresponding length distributions of x-direction and y-direction lengths a and b are shown in Figure 3c,d, according to which we design and simulate the planar metalens with 60 unit cells with x- and y-polarization incidences, respectively. The electric field intensity distributions for x- and y-polarization incidences are shown in Figure 3e,f, respectively. The reflected lights are well focused in the air side except for a little deviation in the focal length (4.95 and 13.32 mm for x- and y-polarization incidences) from the theoretical values. Because the phase shift resulting from the length change cannot cover the full 2π range, we select the adjacent unit cell instead as an approach. The longitudinal and transverse sizes of the focal spots are represented by full width at half maximum (FWHM) of the focal spots along x- and z- directions. For x-polarization incidence, the FWHM values along x- and z- directions are 0.78 and 2.72 mm. For y-polarization incidence, The FWHM values are 1.1 and 9.2 mm, respectively. The focusing efficiencies, defined as the proportion of the incident light energy going to the central focal spot, are calculated to be 41% and 45% for x- and y-polarization incidences. Thus, the metalens with two polarization-controlled focal points is achieved. In practice, a minor inclination of polarization plane is inevitable, which results in a negligible focusing effect for another polarization state.

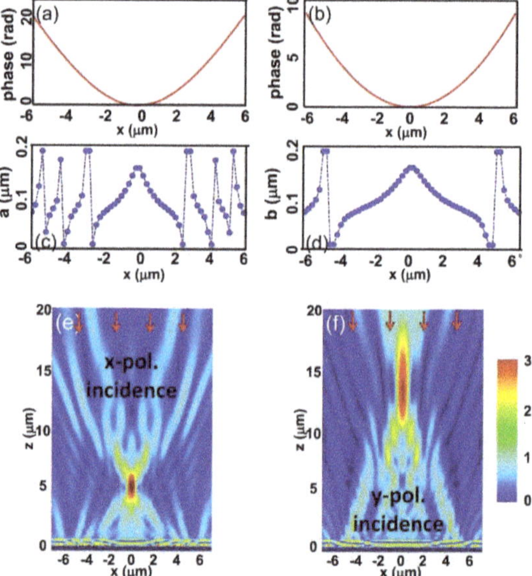

Figure 3. Phase profile to focus incident light for (**a**) x- and (**b**) y-polarization incidences. Corresponding lengths of (**c**) x-direction length a and (**d**) y-direction length b of the rectangle antenna in metalens design. The electric field intensities for (**e**) x- and (**f**) y-polarization incidence, respectively.

To investigate the influence of the x-direction length a and y-direction length b on the focusing effect of y- and x-polarization incidences, we design two metalenses that focus x-polarized light and y-polarized light only. For x-polarization incidence, only the x-direction length a is designed for focusing with a focal length of $F = 5$ μm as shown in Figure 4a, while the y-direction length b is fixed as 100 nm as shown in Figure 4c. For y-polarized light focusing, x-direction length a is fixed as 100 nm as shown in Figure 4b, while the y-direction length b is designed for focusing with a focal length of $F = 15$ μm as shown in Figure 4d. We simulate the two metalenses with x- and y-polarization incidences, and the results are shown in Figure 4e,f, from which we can see that the results are nearly the same as that shown in Figure 3e,f. Therefore, the influence of the x-direction length *a* and y-direction length *b* on the focusing effects of y- and x-polarization incidences are negligible.

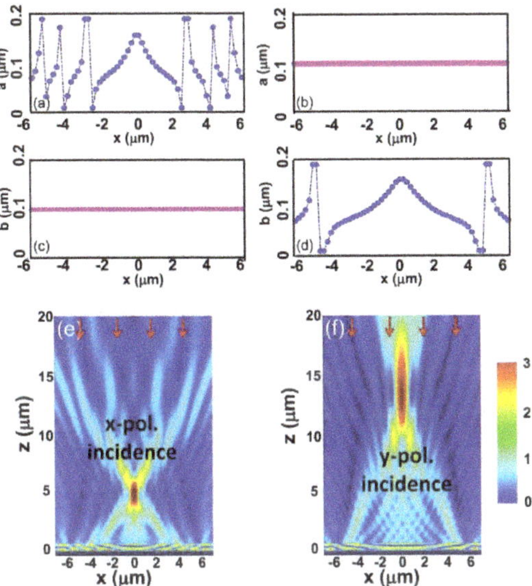

Figure 4. (**a**) X-direction length *a* and (**c**) y-direction length *b* of the metalens for x-focusing only. (**b**) X-direction length *a* and (**d**) y-direction length b of the metalens for y-focusing only. The electric field intensities for (**e**) x- and (**f**) y-polarization incidence, respectively.

In addition, we also investigate the cases of the proposed metalens working in other incident wavelengths. The simulated results are shown in Figure 5. Figure 5a,b show the electric field distributions when illuminated with 750 nm wavelength x- and y-polarized light, respectively. Figure 5c,d show that of 850 nm wavelength incident light. From Figure 5 we can conclude that the proposed metalens can work well within a broadband wavelength that ranges from 750 to 850 nm. In addition, we investigate the focusing effects for other incident wavelengths. The focal lengths for different incident wavelengths are listed in Figure 5e,f, from which we can see that the focal length changes when incident light wavelength is changing, and the focal length is inversely proportional to the wavelength of light.

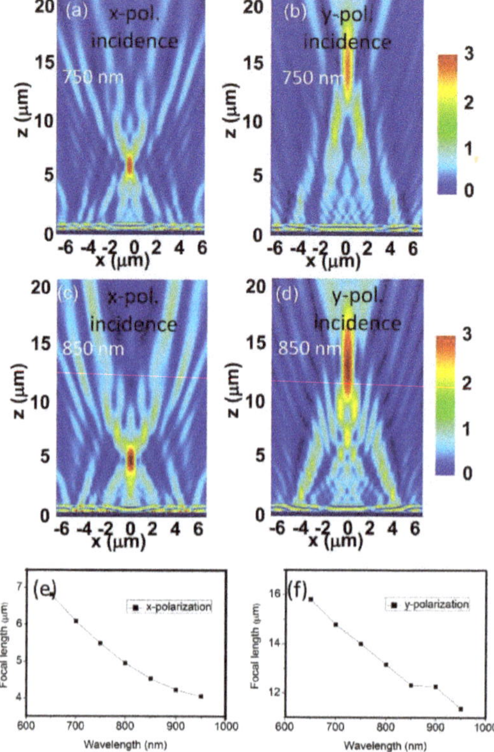

Figure 5. The electric field intensities for (**a**) x- and (**b**) y-polarization incidences with the incident wavelength of 750 nm. The electric field intensities for (**c**) x- and (**d**) y-polarization incidences with the incident wavelength of 850 nm. (**e**,**f**) The focal lengths with different incident wavelengths for x- and y-polarization incidences, respectively.

Furthermore, another function called the beam splitter can be considered in the dual functional metalens design, which manipulates the normal incident light into two opposite oblique paths [35]. To bend the normal incident light to an oblique reflected light, a phase gradient should be introduced along the interface of the structure following the generalized Snell's law [36]. For normal incidence, the reflected angle can be expressed by:

$$\sin(\theta_r) = \frac{\lambda_0}{2\pi n_i} \frac{d\varphi}{dx} \tag{2}$$

where λ_0, x, φ and n_i are the incident wavelength, surface length, phase shift and refractive index of the incident side material, respectively. In our design, the reflection angle of the oblique beams are $\pm30°$, then the calculated phase shift between two adjacent unit is 45°. The phase profile calculated from Equation (2) is shown in Figure 6b. The corresponding y-direction length b distribution is shown in Figure 6d, based on which the normal incident y-polarized light can be split. However, the reflection angle is less than 30°, because the phase shift produced by the gold antennas cannot cover the full 2π range, and we take some approximations in the designing process. Then for x-polarization incidence, the x-direction length a is designed according to Figure 6a,c, which focus the x-polarized incident light with the focal length of $F = 5$ μm. The simulated results of both x- and y-polarization incidences are shown in Figure 6e,f. For x-polarization incidence, the reflected light is well focused at the point of $(x = 0 \, z = 5 \, μm)$, and the y-polarized light is split.

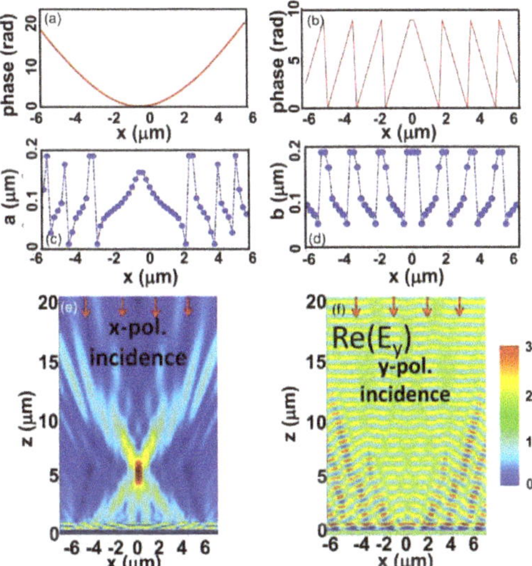

Figure 6. Phase profile (**a**) to focus incident light for x-polarization incidences and (**b**) to split incident light y-polarization incidences. Corresponding lengths of (**c**) x-direction length *a* and (**d**) y-direction length *b* of the rectangle antenna in metalens design. (**e**) The electric field intensities for x-polarization incidence. (**f**) The real part of y-component electric distribution for y-polarization incidence.

The advantage of our results is listed in Table 1 to compare with other works.

Table 1. The compare of this work to previous works.

Literature	Characteristics
Ref. [12,25]	Polarization independent
Ref. [20–24]	X-polarization
This work	Polarization dependent

4. Conclusions

In summary, we proposed a dual functional metalens controlled by polarization state of the incident light. Two functions can be achieved by adjusting the incident light with two polarization states. The proposed metalens is designed according to the approach that the dipole resonance is not influenced by the width of the rectangle gold antenna. A metalens working at 800 nm wavelength with focal lengths of $F = 5$ μm for x-polarization and $F = 15$ μm for y-polarization incidence is designed, which agree well with the two exact metalens that focus x- and y-polarized light only. The proposed metalens works well within a broadband wavelength that ranges from 750 to 850 nm. We also designed a metalens with two functions of focusing and splitting controlled by incident polarization states. Therefore, the proposed metalens can achieve two functions by applying two orthogonal polarized lights.

Author Contributions: Conceptualization, Y.Z. and Y.F.; Methodology, Y.Z.; Software, Y.Z. and Y.F.; Validation, Y.Z. and Y.F.; Formal analysis, Y.Z.; Investigation, B.Y.; Resources, Y.Z. and B.Y.; Data Curation, Y.Z. and B.Y.; Writing-Original Draft Preparation, Y.Z.; Writing-Review and Editing, Z.L. and Y.F.; Supervision, Y.F.; Project Administration, Y.F.; Funding Acquisition, Y.F. All authors have read and agreed to the published version of the manuscript.

Funding: This work is funded by National Natural Science Foundation of China (NSFC) (11474041); National Natural Science Foundation of China (NSFC) (61805025); "111" Project of China (D17017).

Conflicts of Interest: The authors declare no conflict of interest.

References

1. Smith, D.R.; Padilla, W.J.; Vier, D.C.; Nemat-Nasser, S.C.; Schultz, S. Composite Medium with Simultaneously Negative Permeability and Permittivity. *Phys. Rev. Lett.* **2000**, *84*, 4184–4187. [CrossRef]
2. Smith, D.R.; Pendry, J.B.; Wiltshire, M.C.K. Metamaterials and Negative Refractive Index. *Science* **2004**, *305*, 788–792. [CrossRef]
3. Zhou, J.; Dong, J.; Wang, B.; Koschny, T.; Kafesaki, M.; Soukoulis, C.M. Negative refractive index due to chirality. *Phys. Rev. B* **2009**, *79*, 121104. [CrossRef]
4. Ding, J.; Arigong, B.; Ren, H.; Zhou, M.; Shao, J.; Lin, Y.K.; Zhang, H.L. Efficient multiband and broadband cross polarization converters based on slotted L-shaped nanoantennas. *Opt. Express* **2014**, *22*, 29143–29151. [CrossRef]
5. Xu, J.; Li, R.Q.; Qin, J.; Wang, S.Y.; Han, T.C. Ultra-broadband linear polarization converter based on anisotropic metasurface. *Opt. Express* **2018**, *26*, 26235–26241. [CrossRef]
6. Jing, X.F.; Gui, X.C.; Zhou, P.W.; Hong, Z. Physical Explanation of Fabry–Perot Cavity for Broadband Bilayer Metamaterials Polarization Converter. *J. Lightwave Technol.* **2018**, *36*, 2322–2327. [CrossRef]
7. Peng, Y.; Zang, X.; Zhu, Y.; Shi, C.; Chen, L.; Cai, B.; Zhuang, S. Ultra-broadband terahertz perfect absorber by exciting multi-order diffractions in a double-layered grating structure. *Opt. Express* **2015**, *23*, 2032–2039. [CrossRef] [PubMed]
8. Yao, G.; Ling, F.; Yue, J.; Luo, C.; Ji, J.; Yao, J. Dual-band tunable perfect metamaterial absorber in the THz range. *Opt. Express* **2016**, *24*, 1518–1527. [CrossRef] [PubMed]
9. Ning, Y.; Dong, Z.; Si, J.; Deng, X. Tunable polarization-independent coherent perfect absorber based on a metal-graphene nanostructure. *Opt. Express* **2017**, *25*, 32467–32474. [CrossRef]
10. Kildishev, A.V.; Boltasseva, A.; Shalaev, V.M. Planar photonics with metasurfaces. *Science* **2013**, *339*, 1232009. [CrossRef]
11. Yu, N.; Capasso, F. Flat optics with designer metasurfaces. *Nat. Mater.* **2014**, *13*, 139–150. [CrossRef]
12. Wang, W.; Guo, Z.; Li, R.; Zhang, J.; Li, Y.; Liu, Y.; Wang, X.; Qu, S. Plasmonics metalens independent from the incident polarizations. *Opt. Express* **2015**, *23*, 16782–16791. [CrossRef]
13. Ding, P.; Li, Y.; Shao, L.; Tian, X.; Wang, J.; Fan, C. Graphene aperture-based metalens for dynamic focusing of terahertz waves. *Opt. Express* **2018**, *26*, 28038–28050. [CrossRef] [PubMed]
14. Sun, S.; Yang, K.; Wang, C.; Juan, T.; Chen, W.; Liao, C.; He, Q.; Kung, W.; Guo, G.; Zhou, L.; et al. High-Efficiency Broadband Anomalous Reflection by Gradient Meta-Surfaces. *Nano Lett.* **2012**, *12*, 6223–6229. [CrossRef]
15. Luo, L.; Wang, K.; Guo, K.; Shen, F.; Zhang, X.; Yin, Z.; Guo, Z. Tunable manipulation of terahertz wavefront based on graphene nmetasurfaces. *J. Opt.* **2017**, *19*, 115104. [CrossRef]
16. Zhang, Q.; Li, M.; Liao, T.; Cui, X. Design of beam deflector, splitters, wave plates and metalens using photonic elements with dielectric metasurface. *Opt. Commun.* **2018**, *411*, 93–100. [CrossRef]
17. Zhu, Y.; Yuan, W.; Li, W.; Sun, H.; Qi, K.; Yu, Y. TE-polarized design for metallic slit lenses: A way to deep-subwavelength focusing over a broad wavelength range. *Opt. Lett.* **2018**, *43*, 206–209. [CrossRef] [PubMed]
18. Zhang, J.; Guo, Z.; Ge, C.; Wang, W.; Li, R.; Sun, Y.; Shen, F.; Qu, S.; Gao, J. Plasmonic focusing lens based on single-turn nano-pinholes array. *Opt. Express* **2015**, *23*, 17883–17891. [CrossRef] [PubMed]
19. Jia, Y.; Lan, T.; Liu, P.; Li, Z. Polarization-insensitive, high numerical aperture metalens with nanoholes and surface corrugations. *Opt. Commun.* **2018**, *429*, 100–105. [CrossRef]
20. Li, Z.; Yao, K.; Xia, F.; Shen, S.; Tian, J.; Liu, Y. Graphene Plasmonic Metasurfaces to Steer Infrared Light. *Sci. Rep.* **2015**, *5*, 12423. [CrossRef] [PubMed]
21. Zhao, H.; Chen, Z.; Su, F.; Ren, G.; Liu, F.; Yao, J. Terahertz wavefront manipulating by double-layer graphene ribbons metasurface. *Opt. Commun.* **2017**, *402*, 523–526. [CrossRef]
22. Ma, W.; Huang, Z.; Bai, X.; Zhang, P.; Liu, Y. Dual-band light focusing using stacked graphene metasurfaces. *ACS Photonics* **2017**, *4*, 1770–1775. [CrossRef]
23. Yao, W.; Tang, L.; Wang, J.; Ji, C.; Wei, X.; Jiang, Y. Spectrally and spatially tunable terahertz metasurface lens based on graphene surface plasmons. *IEEE Photonics J.* **2018**, *10*, 1–8. [CrossRef]

24. Yin, Z.; Zheng, Q.; Wang, K.; Guo, K.; Shen, F.; Zhou, H.; Sun, Y.; Zhou, Q.; Gao, J.; Luo, L.; et al. Tunable dual-band terahertz metalens based on stacked graphene metasurfaces. *Opt. Commun.* **2018**, *429*, 41–45. [CrossRef]

25. Wang, W.; Guo, Z.; Zhou, K.; Sun, Y.; Shen, F.; Li, Y.; Qu, S.; Liu, S. Polarization-independent longitudinal multi-focusing metalens. *Opt. Express* **2015**, *23*, 29855–29866. [CrossRef] [PubMed]

26. Ji, R.; Chen, K.; Ni, Y.; Hua, Y.; Long, K.; Zhuang, S. Dual-Focuses Metalens for Copolarized and Cross-Polarized Transmission Waves. *Adv. Condens. Matter Phys.* **2018**, *2018*, 2312694. [CrossRef]

27. Bao, Y.; Jiang, Q.; Kang, Y.; Zhu, X.; Fang, Z. Enhanced optical performance of multifocal metalens with conic shapes. *Light Sci. Appl.* **2017**, *6*, e17071. [CrossRef]

28. Bao, Y.; Zu, S.; Liu, W.; Zhou, L.; Zhu, X.; Fang, Z. Revealing the spin optics in conic-shaped metasurfaces. *Phys. Rev. B* **2017**, *95*, 081406(R). [CrossRef]

29. Fan, Q.; Wang, Y.; Liu, M.; Xu, T. High-efficiency, linear-polarization-multiplexing metalens for long-wavelength infrared light. *Opt. Lett.* **2018**, *43*, 6005–6008. [CrossRef]

30. Jiang, Q.; Bao, Y.; Lin, F.; Zhu, X.; Zhang, S.; Fang, Z. Spin-Controlled Integrated Near- and Far-Field Optical Launcher. *Adv. Opt. Mater.* **2018**, *28*, 1705503. [CrossRef]

31. Shao, L.; Premaratne, M.; Zhu, W. Dual-functional coding metasurfaces made of mnisotropic all-dielectric resonators. *IEEE Access* **2019**, *7*, 45716–45722. [CrossRef]

32. Liu, S. Anisotropic coding metamaterials and their powerful manipulation of differently polarized terahertz waves. *Light Sci. Appl.* **2016**, *5*, e16076. [CrossRef] [PubMed]

33. Johnson, P.B.; Christy, R.W. Optical Constants of the Noble Metals. *Phys. Rev. B* **1972**, *6*, 4370. [CrossRef]

34. Zhang, Y.; Zhao, J.; Zhou, J.; Liu, Z.; Fu, Y. Switchable polarization selective terahertz wavefront manipulation in a graphene metasurface. *IEEE Photonics J.* **2019**, *11*, 4600909. [CrossRef]

35. Ramezani, S.A.; Arik, K.; Farajollahi, S.; Khavasi, A.; Kavehvash, Z. Beam manipulating by gate-tunable graphene-based metasurfaces. *Opt. Lett.* **2015**, *40*, 5383–5386.

36. Yu, N.; Genevet, P.; Kats, M.A.; Aieta, F.; Tetienne, J.P.; Capasso, F.; Gaburro, Z. Light Propagation withPhase Discontinuities: Generalized Laws of Reflection and Refraction. *Science* **2011**, *334*, 333–337. [CrossRef] [PubMed]

Article

The Analysis of Resistance to Brittle Cracking of Tungsten Doped TiB$_2$ Coatings Obtained by Magnetron Sputtering

Jerzy Smolik, Joanna Kacprzyńska-Gołacka, Sylwia Sowa * and Artur Piasek

Łukasiewicz Research Networks–Institute for Sustainable Technology, 6/10 Pułaskiego St., 26-600 Radom, Poland; jerzy.smolik@itee.lukasiewicz.gov.pl (J.S.); joanna.kacprzynska-golacka@itee.lukasiewicz.gov.pl (J.K.-G.); artur.piasek@itee.lukasiewicz.gov.pl (A.P.)
* Correspondence: sylwia.sowa@itee.lukasiewicz.gov.pl; Tel.: +48-48-364-44-332

Received: 27 July 2020; Accepted: 19 August 2020; Published: 20 August 2020

Abstract: In this work, the authors present the possibility of characterization of the fracture toughness in mode I (K_{IC}) for TiB$_2$ and TiB$_2$ coatings doped with different concentration of W (3%, 6% and 10%). The Young's modulus, hardness and fracture toughness of this coatings are extracted from nanoindentation experiments. The fracture toughness was evaluated using calculation of crack length measurement. An important observation is that increasing tungsten concentration in the range 0–10% changes the microstructure of the investigated coatings: from columnar structure for TiB$_2$ coating to nano-composite structure for Ti-B-W (10%) coating. It can be concluded that doping with concentration 10 at.% W causes an increase of the fracture toughness for the tested coatings.

Keywords: PVD coatings; nanoindentation; brittle cracking; fracture toughness

1. Introduction

Brittle cracking resistance is one of the most important material features. The fracture toughness in mode I (K_{IC}) is the parameter considered to be a material constant, which determined the material's resistance to brittle cracking. It is independent of the thickness of the material being tested because it refers to a constant state of stress. This makes it possible to assess how much load a structural element containing a crack can carry. The method for determining the K_{IC} factor, which is currently the basic material constant in fracture mechanics, is standardized for solid materials [1,2] and consists of analyzing the process of cracking samples with a properly prepared notch in the three-point bending test. The use of the penetration method to study the fracture toughness K_{IC} was initiated in the 1970s by the Evans and Charles, who observed the relation between the crack lengths, which was generated in the corners of Vickers indenter during the hardness test and the value of K_{IC} [3]. This made it possible to use the microindentation and nanoindentation methods for the mechanical characterization of micro-volume systems, including layers and coatings.

The mechanical characteristics of thin coatings are of great importance in the processes of optimization and development of material solutions for coatings with high tribological efficiency. For mechanical characterization of thin coatings, including determination of their hardness (H), Young's modulus (E) and fracture toughness as the K_{IC} coefficient, nanoindentation has proved to be a very effective technique. Hardness and Young's modulus are determined on the basis of load and displacement curves, while the K_{IC} coefficient can be estimated based on the crack lengths initiated by the indenter.

It should be remembered that, in the case of thin coatings, the results of the brittleness index tests obtained by the nanoindentation method cannot be verified by the impact method, as it is possible for solid materials. Therefore, in the case of solid materials, we can more precisely select the model and range of indenter loads in nanoindentation tests to determine the brittleness of specific materials, e.g., SiC [4], glass, and Al_2O_3 [5]. In the case of thin coatings, we additionally encounter a very large number of material combinations, i.e., chemical composition, phase structure, state of internal stresses resulting from, e.g., the type of substrate or the method of coating deposition. Therefore, it seems very difficult to indicate one model for the determination of the K_{IC} coefficient for all coatings in the current state of knowledge. It has been shown that methods using the nanoindentation method and crack length analysis to determine the brittleness of materials lead to comparable results [6]. This can be a good tool for comparing the brittleness of materials. Importantly, in the case of thin coatings, it is currently difficult to find another method that has as much potential and capabilities in determining the fracture toughness of thin coatings as the nanoindentation method.

The literature analyses show different models for determining the K_{IC} coefficient of thin coatings [7,8], of which two are most often used, as proposed by Anstis's [9] and Laugier [10].

The aim of the research, the results of which are presented in this article, was to propose a method for determining the K_{IC} coefficient of thin PVD coatings using the nanoindentation method. In the article, the Laugier model was used to analyze changes in fracture toughness (K_{IC}) for TiB_2 ceramic coatings doped with tungsten. The coatings were produced by magnetron sputtering in a Direct Current (DC) system. The problem to be solved was the methodology of selecting loads for the Berkovich indenter in the indentation process and the methodology of measuring the crack lengths generated in the coating.

2. Materials and Methods

2.1. Coating Deposition

Tungsten doped TiB_2 coatings were prepared by DC magnetron sputtering method using original magnetron systems made by Łukasiewicz Research Network-Institute for Sustainable Technology in Radom (Ł-ITeE Radom) with a Balzers pump system (Radom, Poland) with two circular magnetrons placed at an angle of 120° to each other. In the deposition process was using two targets made of TiB2 (99.50% purity) and pure-W (99.95% purity) according the procedure present in paper [11]. The diameter of targets was d = 100 mm and thickness g = 7 mm. The TiB_2 and Ti-B-W coatings were deposited in an atmosphere of pure argon (Ar 100%). The parameters of the Ti-B-W coating process are shown in Table 1.

Table 1. Deposition parameters of reference TiB_2 coating and TiB_2 coatings doped with tungsten.

Coating	Atmosphere	Pressure	U_{Bias}	Power of Magnetron		Temperature
		(Pa)	(V)	P_{TiB_2} (W)	P_W (W)	T (°C)
TiB_2				1000	–	
Ti-B-W (1)	Ar 100%	0.5	−50	1000	25	300
Ti-B-W (2)				1000	50	
Ti-B-W (3)				1000	75	

The tested coatings were deposited on to samples made of high-speed steels SW7M with a diameter of 25 mm, thickness 6 mm and surface roughness $R_a \leq 0.05$ μm and on samples of monocrystalline Si (100). The samples were washed with 99.9% pure alcohol before being placed in the process chamber. Before the coating process, the samples were ion-etched in the Ar + plasma. During the coating process, the sample temperature was stabilized at 300 °C using resistance heaters. The Ti-B-W coatings were deposited by changes of the source power in the range 25–75 W. The deposition time for each coating was 1 h. The scheme and view of the magnetron system that we used are shown in Figure 1a–c.

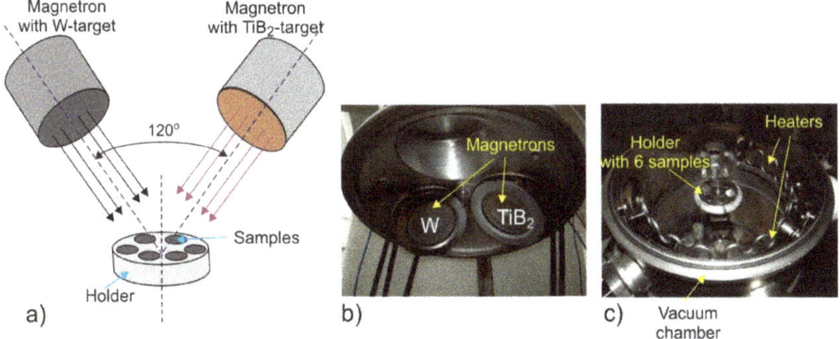

Figure 1. Dual DC magnetron system used to apply TiB$_2$ coatings doped with tungsten: (**a**) scheme; (**b**,**c**) process chamber view.

2.2. Coating Characterization

Samples of monocrystalline Si (100) were used for chemical composition analysis by Wavelength-Dispersive X-Ray Spectroscopy–WDS (Nova NanoSEM 450 with WDS IbeX, Thermo Fisher Scientific, Waltham, MA, USA) localized in AGH (Kraków, Poland). The coatings thickness of brittle fracture cross-section measurements was performed using SEM-Hitachi TM3000 scanning electron microscopy (Radom, Poland).

Samples of SW7M steel with the deposited coatings were subjected to hardness and Young's modulus testing using the Nanoindentation Tester NHT manufactured by Anton Paar with Berkovich diamond indenter (Anton Paar, Ł-ITeE Radom, Poland). For each of the tested coatings, 20 indentations were made in the regime of a contact-depth. The contact-depth of the indenter does not exceed 10% of the coating thickness. The correct indentations were selected and the average hardness values—H and Young's modulus—E, were calculated based on the results obtained, as well as the corresponding standard deviations. The measurement results made it possible to determine the H/E—as the plasticity index or load factor which is responsible for the maximum elastic deflection, when the coating is not destroyed and the H^3/E^2—as a resistance to plastic deformation, which determines the load capacity of the coating.

The surface roughness for all tested coatings were measured by Hommel Tester T1000 produced by JENOPTIK (Ł-ITeE Radom, Poland) by contact method. The mean values of R_a, R_z and R_t parameters were calculated.

2.3. Berkovich Indentation Fracture Toughness Test

SW7M steel samples were also used to study the resistance of brittle cracking using the nanohardness tester CSM with Berkovich diamond indenter. The analysis was carried out in two steps. In the first step, indenter load was selected, when radial cracks were generated in the tested TiB$_2$ and Ti-B-W coatings. For this purpose, 5 indentations were made for each of the tested coatings at different values of the intender normal force, i.e., 50, 100, 200, 300 and 400 mN. The indenter load, for which generated cracks were measurable, was selected for each coating. In the second step, 20 indentations were made for each coating. For each indentation, individual crack length measurements, i.e., (a_n) and (l_n) were made according to the scheme shown in Figure 2. For each coating, based on a group of 20 indentations, the mean values a and l were determined. Then, according to the Laugier model formula (Equation (1)) [10,12], the value of the fracture toughness K_{IC} was determined. Observation and crack lengths measurement in the area of the indentations were carried out using SEM Hitachi TM3000 scanning electron microscopy.

$$K_{IC} = x_v \cdot \left(\frac{a}{l}\right)^{\frac{1}{2}} \cdot \left(\frac{E}{H}\right)^{\frac{2}{3}} \cdot \frac{P}{c^{\frac{3}{2}}} \tag{1}$$

where: K_{IC}—fracture toughness; x_v–indenter geometry factor (for Laugier Equation $x_v = 0.016$); E—Young modulus of coating (GPa); H—hardness of coating (GPa); P–the indentation load (mN); a—the length from the center of the indent to the corner of the indent (μm); $l_{1,2,3}$—the length of the cracks; $l = (l_1 + l_2 + l_3)/3$; $c = l + a$—the total length from the center of the indent to the end of crack.

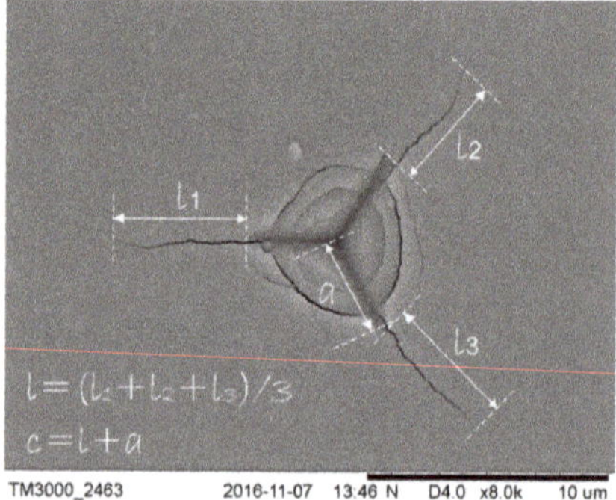

Figure 2. Illustrations of the measurement method of fracture toughness for all investigated indentations.

3. Results

3.1. Coating Characterization

The pure TiB_2 coatings and Ti-B-W coatings doped with tungsten were analyzed after the deposition processes, obtained by the DC magnetron sputtering method, according the parameters presented in Table 1. First, the surface observations and coating thickness measurements were made using the scanning electron microscopy (SEM-Hitachi TM3000) and the surface roughness tests with using the Roughness Hommel Tester. The results of these measurements were necessary to the fracture toughness test, which was planned in the next step. Figure 3 shows the results of surface and cross-sections observations for the tested TiB_2 and Ti-B-W coatings. All tested coatings are characterized by high surface smoothness and good coherence, which are free of cracks and defects. The thickness of all investigated coatings was in the range 1.2–1.3 μm.

The prepared coatings were subjected to chemical composition analysis using the WDS method. The obtained results (Table 2) showed that the tungsten concentration doped into the TiB_2 coating increased with increasing tungsten power and it is range 3%–10%. The Ti/B or B/(Ti+W) ratios were determined adequately in the TiB_2 and Ti-B-W coatings.

Table 2. Parameters for investigated TiB_2 and Ti-B-W coatings.

Coating	Thickness g (μm)	Roughness $R_a/R_z/R_t$ (μm)	Hardness H (GPa)	Young's Modulus E (GP)	Plasticity Index H/E	Resistance to Plastic Deformation H^3/E^2
TiB_2	1.0	0.03/0.22/0.55	34.0 ± 2	405 ± 5	0.075	0.239
Ti-B-W (3%)	1.1	0.01/0.21/0.41	35.5 ± 2	415 ± 10	0.085	0.259
Ti-B-W (6%)	1.2	0.01/0.29/0.56	37.0 ± 2	425 ± 7	0.087	0.280
Ti-B-W (10%)	1.3	0.02/0.13/0.29	38.0 ± 3	435 ± 5	0.087	0.289

The hardness and Young's modulus measurements were carried out without exceeding the depth of 10% of the coating thickness, i.e., max 100 nm. Examples of changes in the force acting on the Berkovich indenter in the case of testing the hardness and Young's modulus of the TiB_2 coating and the view of the indentation. Based on the obtained results, the plasticity index as H/E and the resistance to plastic deformation as H^3/E^2 of investigated coatings were determined. All parameters of the investigated TiB_2 and Ti-B-W coatings are shown in Table 3.

Table 3. The results of investigations of the chemical composition for TiB$_2$ and Ti-W-B coatings, which were made using the WDS method.

Coating	Chemical Composition (at.%)			B/(Ti + W)
	Ti	W	B	
TiB$_2$	31	–	69	2.2
Ti-B-W (1)	22	3	75	3.0
Ti-B-W (2)	22	6	72	2.5
Ti-B-W (3)	22	10	68	2.1

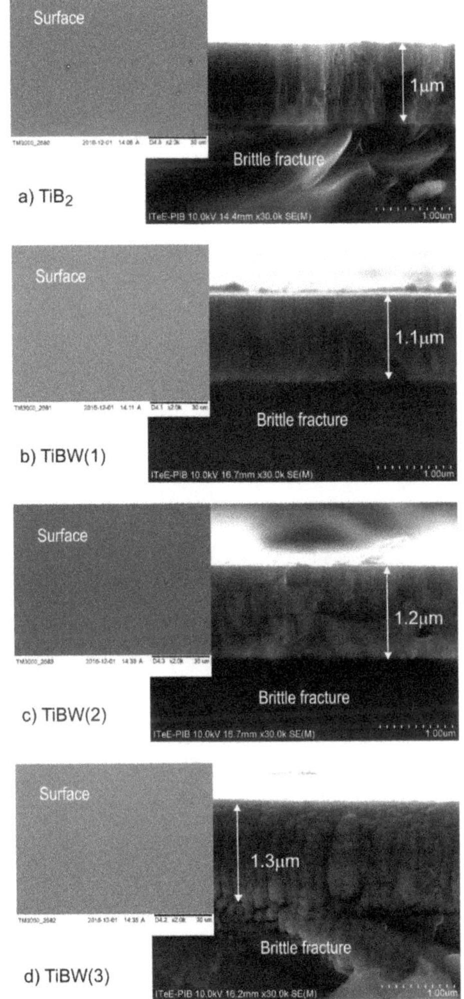

Figure 3. SEM micrographs from the surface (on the left section) and from the brittle fracture cross-section (on the right section) with coating thickness measurements for coatings obtained with different power of tungsten sputtering (P_W in Table 1): (**a**) TiB$_2$: P_W = 0 W; (**b**) Ti-B-W(1): P_W = 25 W; (**c**) Ti-B-W(2): P_W = 50 W; (**d**) Ti-B-W(3): P_W = 75 W.

3.2. Determination of the Fracture Toughness K_{IC}

For measured the sizes of the crack lengths emerging from their corners after indentation with different forces 50, 100, 200, 300 and 400 mN, all the generated cracks were imaged using a scanning electron microscopy (SEM-Hitachi TM3000). The images of indentations for different types of coatings are shown in Figure 4. Observations for the indentations made with the different loads showed, that the cracks for coatings such as TiB_2, Ti-B-W (3%), Ti-B-W (6%), which were generated in the corners of the indentation, were clearly visible and well measurable already at the indenter load of 200 mN (Figure 4a–c). On the other hand, the cracks for coating Ti-B-W (10%) were visible and well measurable only with indenter load of 400 mN (Figure 4d). Such intender loading values were accepted for testing of the fracture toughness K_{IC} for selected coatings.

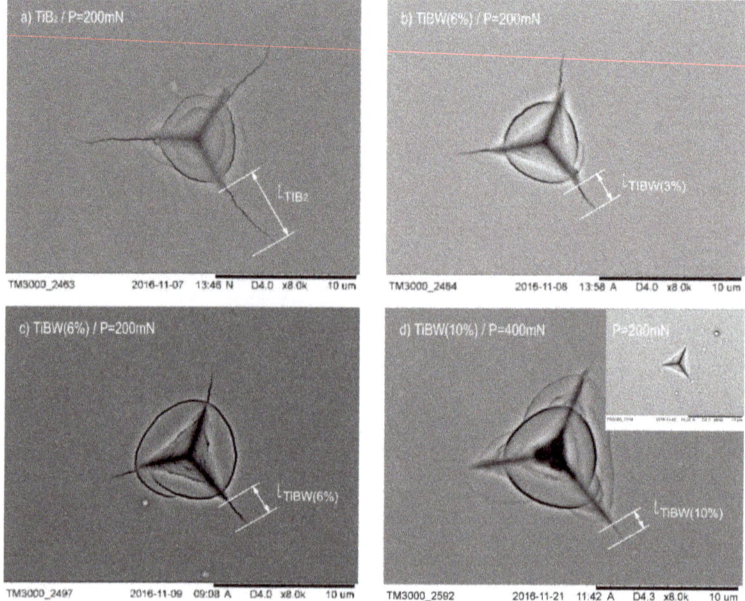

Figure 4. The indentations for selected TiB_2 and Ti-B-W coatings after testing with different values of indenter loading force: (**a**) TiB_2: 200 Mn; (**b**) Ti-B-W (3%): 200 mN; (**c**) Ti-B-W (6%): 200 mN; (**d**) Ti-B-W (10%): 400 mN.

According to the adopted methodology (present in Section 2.1), 20 indentations were made for each coating with the selected indenter load (P), i.e., for the coatings: TiB_2, Ti-B-W (3%), Ti-B-W (6%) was P = 200 mN, while for the coating Ti-B-W (10%) was P = 400 mN. The penetration depth was respectively h_{200mN} = 1000 nm and h_{400mN} = 1600 nm. For each indention, we measured the edge lengths for the indentation a_n and the crack lengths l_{n1}, l_{n2}, l_{n3}, where n is the number of indentation. For each indentation, we determined the average values of l_n according with equation $l_n = (l_{n1} + l_{n2} + l_{n3})/3$. Then, for each tested coating, the average values of l and a were determined for the whole series of 20 indentations in accordance with relation Equations (2) and (3). Figure 5a,b shows an example of series of indentations made for the Ti-B-W (6%) coating and the results of the indentation $n = 1$.

$$a = (a_{n=1} + a_{n=2} + \cdots + a_{n=20})/20 \tag{2}$$

$$l = (l_{n=1} + l_{n=2} + \cdots + l_{n=20})/20 \tag{3}$$

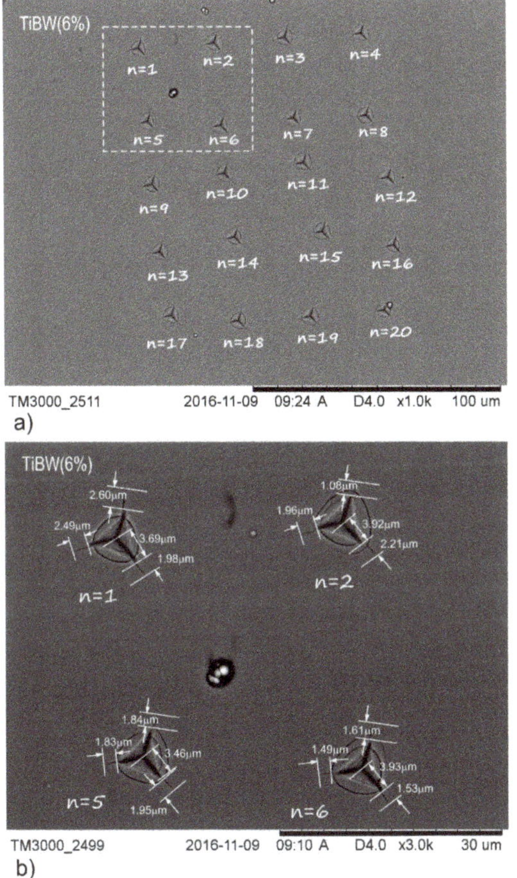

Figure 5. SEM images of groups indentations used for fracture toughness analysis (K_{IC}) for Ti-B-W (6%) coating: (**a**) series of 20 indentations with a Berkovich indenter at a load of P = 200 mN; (**b**) dimensional analysis of crack lengths for different indentations where n = 1, 2, 5, 6.

Based on the material tests and measurements of the crack lengths in the corners of the indentations made of the Berkovich indenter, Table 4 summarizes the parameters that are needed to determine the K_{IC} for the tested coatings, i.e., TiB$_2$, Ti-B-W (3%), Ti-B-W (6%) and Ti-B-W (10%). The value of the coefficient (x_v) in the Lougier model depends on the geometry of the indenter, which was used. According to the analysis carried out by N. Cuadrado et al. [13], for materials such as SiC, Si and soda-lime glass, in the case of Berkovich's geometry the value of coefficient (x_v) = 0.022 ± 0.001.

Table 4. Results of calculation of fracture toughness (K_{IC}), hardness and Young modulus for all investigated coatings.

Coating	x_v	P (mN)	H (GPa)	E (GPa)	a (µm)	l (µm)	c (µm)	$K_{IC}=x_v\cdot\left(\frac{a}{l}\right)^{1/2}\cdot\left(\frac{E}{H}\right)^{2/3}\cdot\frac{P}{c^{3/2}}$
TiB$_2$	0.022	200	34.0 ± 2	405 ± 5	4.20	5.32	9.52	$K_{IC\ [TiB_2]}$ = 0.67
Ti-B-W (3%)	0.022	200	35.5 ± 2	415 ± 10	3.78	2.42	6.20	$K_{IC\ [TiBW\ (3\%)]}$ = 1.84
Ti-B-W (6%)	0.022	200	37.0 ± 2	425 ± 7	3.64	2.10	5.74	$K_{IC\ [TiBW\ (6\%)]}$ = 2.16
Ti-B-W (10%)	0.022	400	38.0 ± 3	435 ± 5	6.20	1.20	4.83	$K_{IC\ [TiBW\ (10\%)]}$ = 4.98

The analysis carried out according to the Laugier model showed that, as a result of doping the TiB_2 coating with tungsten, there is a significant increase in its fracture toughness. When the tungsten concentration increases up to 10%, the cracks that were generated in the corners of the indentation made with Berkovich indenter become shorter. For concentrations of 3%–6% W, the fracture toughness of Ti-B-W coatings achieves a value comparable to K_{IC} suitable for TiN [14] and CrN [15] coatings. The fracture toughness of Ti-B-W with 10 at.% W is $K_{IC\,[TiBW\,(10\%)]} = 4.98$ and is nearly 7.5 times higher than for the TiB_2 coating $K_{IC\,[TiB_2]} = 0.67$.

4. Discussion

In the article, the authors presented how the changes in the fracture toughness K_{IC} for TiB_2 and Ti-B-W coatings depend on the atomic % tungsten concentration (at.% W), as shown in Figure 6a. Observations of the brittle fracture cross-section showed that the TiB_2 coating has a column structure, where the grain diameter is ≈ 100 nm (Figure 3a). The brittle fracture cross-section in the TiB_2 coating has a clear intergranular character. In Figure 6b, the authors presented a scheme of the microstructure of the TiB_2 coating, where the direction of cracking is parallel to the direction of growth of pillar grains (perpendicular to the surface of the coating). The presented diagram explains the high brittleness of the TiB_2 coating and the low value of the fracture toughness K_{IC} ($K_{IC\,[TiB2]} = 0.67$).

Doping the TiB_2 coating with 3% tungsten reduces column grains to a diameter of ≈ 30 nm (Figure 3b). Figure 6c proposes a diagram of the microstructure of the Ti-B-W (3%) coating, where grain refinement does not change the cracking mechanism, which still works mainly in a direction perpendicular to the surface. However, increasing the hardness and Young's modulus of the Ti-B-W (3%) coating, and consequently also increasing the plasticity index of coating H/E and resistance to plastic deformation of coating H^3/E^2, results in increased resistance to brittle cracking ($K_{IC\,[TiBW\,(3\%)]} = 1.84$).

For Ti-B-W coatings, which contain 6% tungsten, the column structure with diameter of grains 30 nm also dominates (Figure 3c). The cracking mechanism is noticeably changed, where the directions of parallel and perpendicular cracking to the surface of the coating are equivalent. At the brittle fracture cross-section, the columnar grains were observed, which cracked in a direction parallel to the surface of the coating. In Figure 6d, the authors proposed a scheme of the Ti-B-W (6%) coating microstructure, where columnar grains change their structure and growth directions due to tungsten segregation and the possibility of new phases appearing, e.g., WB_4. The occurrence of local differences in the microstructure and phase structure justifies the increase in the hardness of the TiBW (6%) coating (H_{TiBW} (6%) = 37 GPa) and the possibility of cracking energy dissipation by changing the direction of cracking and its separation into several others.

In the Ti-B-W coating, obtained by the TiB_2 coating doped with 10% tungsten, one can observe a compact columnar structure with the individual columns, which are agglomerates of equiaxed grains with a diameter ≈ 100 nm. This allows us to make conclusions regarding the possible occurrence of nano-composite microstructure in Ti-B-W (10%) coatings, which was confirmed by the results of the research presented by Sobol et al. [16]. In Figure 6e, the authors presented a scheme of the Ti-B-W (10%) coating microstructure, where the cracking process is not oriented, but indicates the possibility of energy dissipation during cracking. In addition, the doping 10% of tungsten in the structure of the TiB_2 coating results in a large increase in hardness and Young's modulus, i.e., $H_{TiB2} = 34$ GPa $\rightarrow$ $H_{TiBW\,(10\%)} = 38$ GPa and $E_{TiB2} = 405$ GPa $\rightarrow E_{TiBW\,(10\%)} = 435$ GPa, and the increase plasticity index H/E of coating and resistance to plastic deformation H^3/E^2 of Ti-B-W (10%) coatings, which are respectively: $H/E_{TiBW\,(10\%)} = 0.087$ and $H^3/E^2_{TiBW\,(10\%)} = 0.289$. As a result, the fracture toughness is significantly increased: $K_{IC\,TiBW\,(10\%)} = 4.98$.

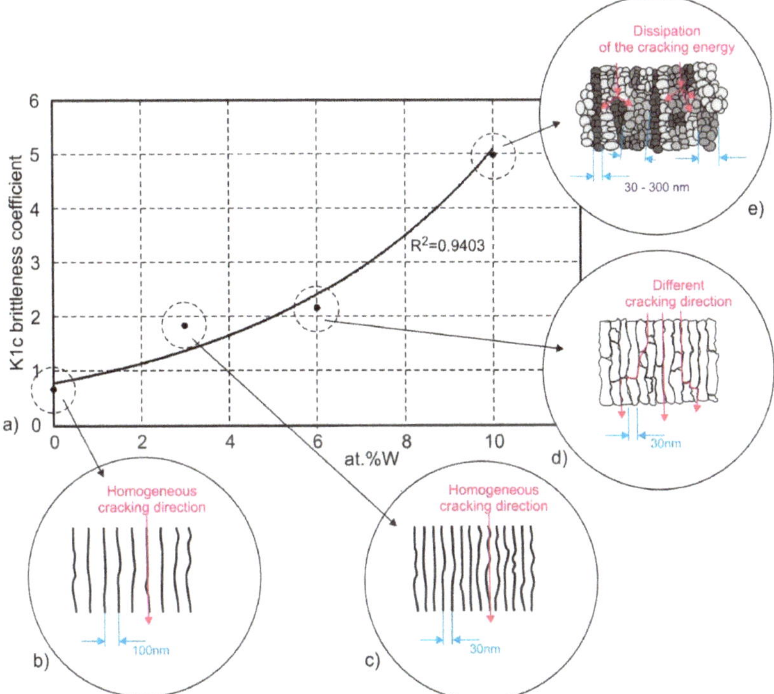

Figure 6. The results of the fracture toughness K_{IC} analysis for TiB_2 and Ti-B-W coatings: (**a**) changes of values of K_{IC} for TiB_2 and Ti-B-W coatings depending on the tungsten concentration (at. % W); (**b**) scheme of microstructure TiB_2 coating; (**c**) scheme of microstructure Ti-B-W (3%) coating; (**d**) scheme of microstructure Ti-B-W(6%) coating; (**e**) scheme of microstructure Ti-B-W (10%) coating.

5. Conclusions

The article demonstrated that the brittleness of thin TiB_2 ceramic coatings can be effectively improved by tungsten doping. Tungsten concentration in the range of 0–10% can significantly changes the microstructure of coatings, from a typical columnar structure with a grain diameter of about 100 nm for TiB_2, to a clear nano-composite structure for Ti-B-W (10%). Analysis of the tested TiB_2 and Ti-B-W coatings, including: brittle fracture cross-section, changes in hardness (H) and Young's modulus (E) and plasticity index H/E and resistance to plastic deformation H^3/E^2, enabled the development of microstructure diagrams for TiB_2 coatings and Ti-B-W coatings with different tungsten concentration. Observations of brittle fracture cross-section show that, in the case of doping TiB_2 coatings with tungsten at an amount of 10%, the microstructure is fragile and the nano-composite structure is created. Then it is possible to change the direction of cracking many times and to disperse it into several other directions. The energy of a single crack is reduced, which often results in the disappearance of cracks. As a result, in such a coating we observed a significant increase in values of fracture toughness K_{IC}.

To assess the fragility of the tested coatings, the authors used the Berkovich indenter induction method and the calculation of the fracture toughness K_{IC} according to Laugier model. Doping of TiB_2 coating show that 10% tungsten causes a more than 7-times increase in the fracture toughness K_{IC} from $K_{IC\,[TiB2]} = 0.67$ to $K_{IC\,[TiBW\,(10\%)]} = 4.98$.

At the same time, Ti-B-W (10%) coatings are characterized by greater hardness and comparable surface roughness compared to TiB_2 coating, which makes them coatings with a very high application potential.

Author Contributions: Conceptualization, J.S.; methodology, J.S., J.K.-G., and S.S.; investigation, S.S. and A.P.; data curation, J.S, J.K.-G., and S.S.; writing—original draft preparation, J.S.; writing—review and editing, S.S. All authors have read and agreed to the published version of the manuscript.

Funding: This research received no external funding.

Conflicts of Interest: The authors declare no conflict of interest.

References

1. *ASTM G40-15 Standard Terminology Relating to Wear and Erosion*; ASTM: West Conshohocken, PA, USA, 1987; Volume 3, pp. 243–250.
2. Zhu, X.-K.; Joyce, J.A. Review of fracture toughness (G, K, J, CTOD, CTOA) testing and standardization. *Eng. Fract. Mech.* **2012**, *85*, 1–46. [CrossRef]
3. Lawn, B.R.; Evans, A.G.; Marshall, D.B. Elastic/plastic indentation damage in ceramics: The median/radial crack system. *J. Am. Ceram. Soc.* **1980**, *63*, 574–581. [CrossRef]
4. Datye, A.; Schwarz, H.T.; Lin, U.D. Fracture toughness evaluation and plastic behavior law of a single crystal silicon carbide by nanoindentation. *Ceramics* **2018**, *1*, 198–210. [CrossRef]
5. Lee, K.; Marimuthu, K.P.; Lee, J.H.; Rickhey, F.; Han, J.; Lee, H. Determination of crack-free mechanical properties of brittle materials via single nanoindentation. *Int. J. Solids Struct.* **2020**, *191*, 8–25. [CrossRef]
6. Hervas, I.; Montagne, A.; Van Gorp, A.; Bentoumi, M.; Thuault, A.; Iost, A. Fracture toughness of glasses and hydroxyapatite: A comparative study of 7 methods by using Vickers indenter. *Ceram. Int.* **2016**, *42*, 12740–12750. [CrossRef]
7. Feng, Y.; Zhang, T. Determination of fracture toughness of brittle materials by indentation. *Acta Mech. Solida Sin.* **2015**, *28*, 221–234. [CrossRef]
8. Sebastiani, M.; Johanns, K.E.; Herbert, E.G.; Pharr, G.M. Measurement of fracture toughness by nanoindentation methods: Recent advances and future challenges. *Curr. Opin. Solid State Mater. Sci.* **2015**, *19*, 324–333. [CrossRef]
9. Anstis, G.R.; Chantikul, P.; Lawn, B.R.; Marshall, D.B. A critical evaluation of indentation techniques for measuring fracture toughness: I, direct crack measurements. *J. Am. Ceram. Soc.* **1981**, *64*, 533–538. [CrossRef]
10. Laugier, M.T. New formula for indentation toughness in ceramics. *J. Mater. Sci. Lett.* **1987**, *6*, 355–356. [CrossRef]
11. Smolik, J.; Mazurkiewicz, A.; Garbacz, H.; Kopia, A. Tungsten doped TiB_2 coatings obtained by magnetron sputtering. *J. Mach. Constr. Maint.* **2018**, *4*, 27–32.
12. Fabijanic, T.A.; Coric, D.; Musa, M.Š.; Sakoman, M. Vickers indentation fracture toughness of near-nano and nanostructured WC-Co cemented carbides. *Metals* **2017**, *7*, 1–16.
13. Cuadrado, N.; Casellas, D.; Anglada, M.; Jiménez-Piqué, E. Evaluation of fracture toughness of small volumes by means of cube-corner nanoindentation. *Scr. Mater.* **2012**, *66*, 670–673. [CrossRef]
14. Zhang, L.; Yang, Y.; Pang, X.; Gao, K.; Volinsky, A.A. Microstructure, residual stress, and fracture of sputtered TiN films. *Surf. Coat. Technol.* **2013**, *224*, 120–125. [CrossRef]
15. Daniel, R.; Meindlhumer, M.; Zalesak, J.; Sartory, B.; Zeilinger, A.; Mitterer, C.; Keckes, J. Fracture toughness enhancement of brittle nanostructured materials by spatial heterogeneity: A micromechanical proof for CrN/Cr and TiN/SiO_x multilayers. *Mater. Des.* **2016**, *104*, 227–234. [CrossRef]
16. Sobol, O.V.; Dub, S.N.; Pogrebnjak, A.D.; Mygushchenko, R.P.; Postelnyk, A.A.; Zvyagolsky, A.V.; Tolmachova, G.N. The effect of low titanium content on the phase composition, structure, and mechanical properties of magnetron sputtered WB2-TiB2 films. *Thin Solid Films* **2018**, *662*, 137–144. [CrossRef]

Article

Improvement of the Interfacial Fatigue Strength and Milling Behavior of Diamond Coated Tools via Appropriate Annealing

Georgios Skordaris [1,*], Tilemachos Kotsanis [1], Apostolos Boumpakis [1] and Fani Stergioudi [2]

[1] Laboratory for Machine Tools and Manufacturing Engineering, Mechanical Engineering Department, Aristotle University of Thessaloniki, 54124 Thessaloniki, Greece; tilemahoskotsanis@gmail.com (T.K.); ampoumpak@auth.gr (A.B.)

[2] Physical Metallurgy Laboratory, Mechanical Engineering Department, Aristotle University of Thessaloniki, 54124 Thessaloniki, Greece; fstergio@auth.gr

* Correspondence: skordaris@meng.auth.gr; Tel.: +30-2310-996-027

Received: 29 July 2020; Accepted: 22 August 2020; Published: 25 August 2020

Abstract: This article deals with the potential to reduce the amount of the residual stresses in the diamond films on cemented carbide inserts for improving their effective interfacial fatigue strength and thus their wear resistance. In this context, nano-crystalline diamond coatings (NCD) were deposited on cemented carbide inserts. A portion of these coated tools were annealed in vacuum for decreasing the amount of residual stresses in the film structure. The annealing temperature was appropriately selected for keeping the substrate strength properties invariable after the coating annealing. Inclined impact tests at ambient temperature on the untreated and heat-treated diamond coated tools were conducted for evaluating their effective interfacial fatigue strength. Depending upon the impact load, after a certain number of impacts, damages in the film-substrate interface develop, resulting in coating detachment and lifting. Via appropriate FEM (Finite Element Method)-evaluation of the impact imprints, the residual stresses in the diamond film structure were determined. Milling experiments were conducted for evaluating the cutting performance of the coated tools using aluminum foam as workpiece material. A correlation between the interfacial fatigue strength of diamond coatings and their residual stresses affected by annealings contributed to the explanation of the attained cutting results.

Keywords: diamond coatings; residual stresses; interfacial fatigue strength; annealing; milling

1. Introduction

Diamond coatings deposited on cemented carbide tools are widely used in machining of non-ferrous materials like aluminum alloys, composite and so forth [1–5]. A key issue for ensuring a sufficient diamond coated tool life, especially in milling, is the fatigue strength of diamond coating-substrate interface [6–8]. For enhancing the diamond film bonding on its substrate and thus the film adhesion, selective chemical Co-etching is carried out on hard metal tools for decreasing superficially the non-adhesive cobalt [1,4]. Moreover, the application of interlayer materials was introduced as an effective technology for enhancing the film adhesion and nucleation densities of diamond coatings on various substrates [4,6,9–11].

In the present paper, the potential to change the amount of residual stresses in the diamond film structure via appropriate heat treatment, thus increasing the effective interfacial fatigue strength as well as the wear resistance of diamond coated tools in milling was investigated. Nano-crystalline diamond (NCD) coatings deposited on cemented carbide tools are characterized by high residual stresses. The high level of residual stresses in the diamond film structure are attributed to epitaxial crystal differences and thermal expansion coefficients mismatch of the diamond coating and its cemented-carbide substrate [12–14]. In this context, annealings were carried out using the NCD coated tools for attaining a relaxation of the residual stresses [15]. Moreover, Raman spectra were applied to study the crystallinity of the as deposited as well as annealed diamond coatings [5,16]. Nano-indentations coupled with appropriate finite element method (FEM)-simulations were applied for characterizing the substrate mechanical properties at ambient and at annealing temperature [17]. As a result, potential change of the wear behavior of diamond coated tools due to annealing can be attributed only to the relaxation of the film residual stresses.

The effective interfacial fatigue strength of the as deposited and heat-treated NCD coated tools was investigated by conducting inclined impact tests at ambient temperature and at various loads and number of impacts. Under certain impact conditions, the conduct of the inclined impact test results in the development of interfacial fatigue micro-cracks and damages. As a consequence, a release of the residual stresses in the diamond film structure takes place leading to bulge formation [7]. Via appropriate FEM-evaluation of the impact imprints, the residual stresses in the examined diamond coatings were calculated [18]. Moreover, milling experiments were carried out for evaluating the cutting performance of the prepared NCD coated inserts using aluminum foam as workpiece material. Based on the achieved results, optimum annealing conditions can be determined for improving the wear resistance of the NCD coated tools in milling.

2. Experimental and Computational Details

Nano-composite diamond coatings (NCD) were deposited on cemented carbide inserts of HW-K05 SPGN 120308 specifications via hot filament method using a CC800/9Dia Cemecon coating machine (Würselen, Germany). The applied temperature of the filament and substrate during the deposition process were 2000 °C and 900 °C, respectively. Moreover, a cooling process took place after the coating deposition lasting for 9 h. The coating thickness amounts to approximately 5 μm. For achieving a sufficient film adhesion, cobalt-etching procedures were conducted prior to the coating deposition [6]. A portion of the produced diamond inserts was annealed in vacuum for a duration of 3 h and 10 h. The annealing temperature was set equal to 400 °C. In this way, three groups of diamond coated inserts were created possessing different annealing durations.

For characterizing the crystallinity of the as deposited and annealed at different conditions diamond coatings, Raman spectroscopy was carried out using a LabRAM HR spectrometer (HORIBA FRANCE SAS). The recorded spectra are presented in Figure 1. The as deposited as well as the annealed diamond coatings appear a narrow peak at 1332 cm^{-1}. This peak is associated with a diamond coating characterized by high crystalline quality [16]. Moreover, the captured spectra for all investigated coating cases are similar. Thus, it can be stated that the effect of the temperature up to 400 °C on the crystallinity of diamond coatings is negligible.

For evaluating the substrate strength properties before and after annealing at 400 °C, nanoindentations were conducted by a FISCHERSCOPE H100 device (Helmut Fischer GmbH, Sindelfingen, Germany) [17]. The fatigue strength of the coating–substrate interface was assessed by inclined impact tests at various loads and cycles using an impact tester designed and manufactured by the Laboratory for Machine Tools and Manufacturing Engineering of the Aristotle University of Thessaloniki in conjunction with the company Impact-BZ [19]. The load inclination angle was 15° and the ball diameter was 5 mm (see Figure 2a). The applied time-dependent force signal is shown in Figure 2b. The 3D-measuement facilities of the confocal microscope μSURF of NANOFOCUS AG (Oberhausen, Germany) were used for evaluating the impact imprints. The milling experiments were conducted without coolant or lubricant for attaining a more intense wear evolution. The workpiece material was an aluminum foam. Based on standard metallographic techniques, the various hard phases of the workpiece material were detected and described in Figure 2c. It has to be pointed out that the coated cutting edge is loaded by high dynamic loads during milling owing to the structure of workpiece material.

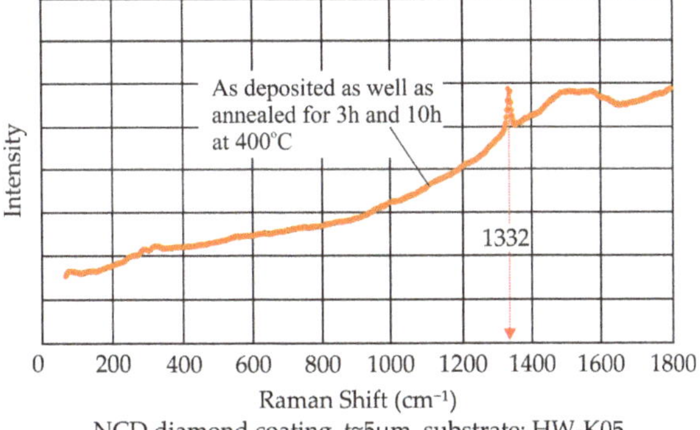

Figure 1. Raman spectra of the as deposited as well as the annealed at different conditions diamond coatings for investigating their crystallinity.

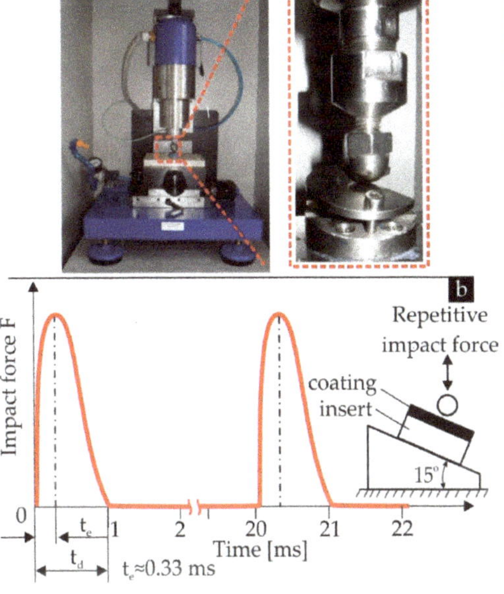

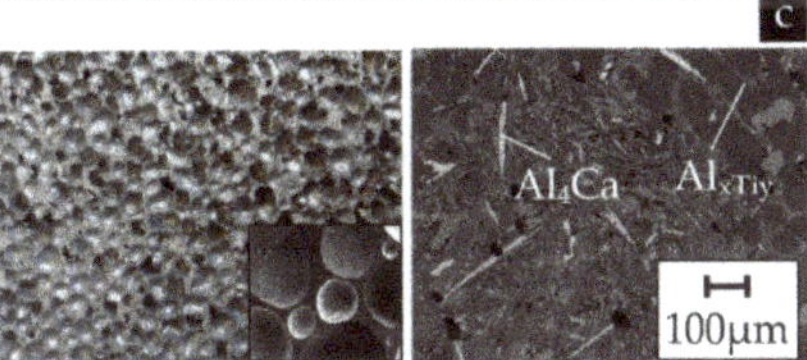

- Eutectic structure (laminate) Al with Al_4Ca.
- The white color (which are mainly elongated) corresponds to $Al_{13}Fe_4$ phase which is orthorhombic phase.
- The gray color corresponds to Al_xTi_y phases (cubic and others hexagonal)
- The foam is oxide-free.

Workpiece material: Aluminium foam
Cemented carbide inserts of HW-K05
SPGN120308 ISO

Figure 2. (**a**) The device for conducting inclined impact tests; (**b**) the applied force signals during the inclined impact test; (**c**) characteristic micro-graphs of the aluminum foam used in milling experiments.

An axisymmetric FEM-model was employed using ANSYS 18 software for determining the thermal residual stresses in the diamond film structure, as shown in Figure 3a and described analytically in Reference [18]. The applied FEM-algorithm (FEM-TRS algorithm) enables the calculation of thermal compressive stresses in the NCD structure during cooling from the deposition temperature to the ambient one taking into account the different thermal expansion coefficients of the diamond coating and the cemented carbide substrate.

Moreover, the maximum residual stresses in the diamond coating structures before and after annealing were determined by an appropriate FEM-model using ANSYS software, as shown in Figure 3b and described analytically in Reference [18]. By using the experimental data of the diameter of detached film region in each examined coating case after inclined impact test, the goal of the applied FEM-DRS (Finite Element Method-Distributed Resource Schedule) algorithm is to describe the geometry of the formed film bulges [18]. In this context, a pressure is applied to the FEM-model elements until the deformation height at the most elevated central bulge region to be equal with the relevant measured film bulge maximum height. Thus, the overall compressive residual stresses in the diamond film structure can be determined.

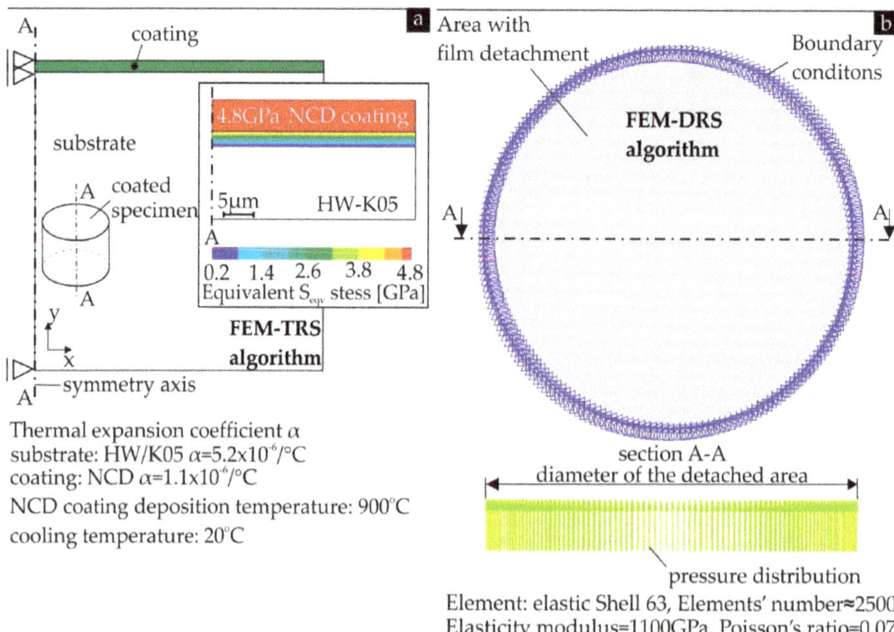

Figure 3. The applied finite element method (FEM) model for calculating (**a**) the thermal residual stresses and (**b**) the overall residual stresses in film structure after the diamond coating deposition and its annealing.

3. Results and Discussion

3.1. Mechanical Properties of Cemented Carbide Substrates before and after Annealing

Since the goal of this work was to improve the effective interfacial fatigue strength of diamond coated tools and thus their wear behavior in milling via appropriate annealing, it was necessary to ensure that the effect of the applied annealing process on substrate mechanical properties is negligible. In this way, uncoated HW-K05 cemented carbide substrates were annealed in vacuum at a temperature of 400 °C corresponding to the related one during the annealing of the diamond coated tools. Herewith, the effect of the annealing temperature on the substrate strength properties was recorded by conducting nanoindentations at a maximum indentation load of 50 mN. In order the effect of the specimen roughness on the results accuracy to be nullified, 30 measurements per nanoindentation were conducted [17]. The related load-displacement diagrams for both examined cases are shown in Figure 4a. According to the obtained results, the load-displacement curve of cemented carbide subjected to an annealing at 400 °C is similar to the corresponding one of an untreated substrate. By appropriate FEM-evaluation of the nanoindentation results, the mechanical characteristics for both cemented carbide inserts are shown in Figure 4b [17]. Based on the attained results, it can be stated that potential change of the wear behavior of diamond coated tools via annealing can be attributed only to the relaxation of the film residual stresses and not to substrate effects.

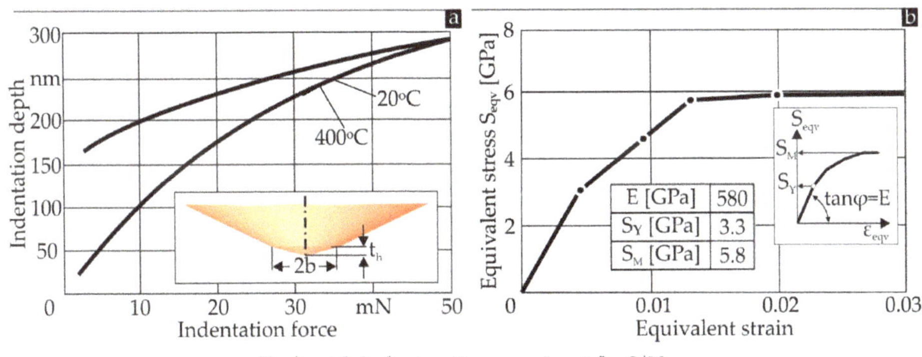

Berkovich indenter, tip geometry: t_h/b= 2/89 nm
NCD coating/ HW-K05 substrate, t≈5 μm

Figure 4. (**a**) Nanoindentation load-displacement curves; (**b**) Calculated stress-strain curves of HW-K05 cemented carbide inserts before and after annealing at 400 °C.

3.2. Residual Stresses in NCD Coatings before and after Their Annealings

Annealings were conducted on the diamond coated tools in vacuum for 3 h and 10 h at a temperature of 400 °C. Taking into account the dominant effect of the strength properties of the cemented carbide substrates on the capability of coated tools to withstand the dynamic loads in interrupted cutting processes, the annealing temperature of 400 °C was intentionally selected. At the temperature of 400 °C, the HW-K05 substrate strength properties remain invariable, as also described in the Section 3.1 [20]. In the first step of the conducted investigations, it was necessary to determine the developed residual stresses in the film structure before and after annealing. Based on an appropriate axisymmetric FEM-model described in Figure 3a (FEM-TRS algorithm), the developed thermal stresses due to the different thermal expansion coefficients of the diamond coating and cemented carbide substrate after cooling from the deposition temperature to the ambient one were calculated and amount to approximately 4.8 GPa. These compressive stresses remain constant in all investigated annealing cases. The magnitude of the thermal residual stresses depends only on the operational temperature of the diamond coated tools and the coating deposition temperature. In all cases, the inclined impact tests for evaluating the interfacial fatigue strength of diamond coated tools were conducted at ambient temperature. Moreover, by using a further FEM-model described in Figure 3b (FEM-DRS algorithm), the overall amount of residual stresses in the film structure were calculated. This FEM-model, as already described in Section 2, simulates the film lifting due to the fatigue failure of its interface at ambient temperature and the residual stress release after the conduct of inclined impact test considering the diameter and height of the formed bulges. The diameter of each detached region after the inclined impact test corresponds to a certain film bulge height depending on the amount of the residual stresses in the diamond coating structure [21].

A characteristic example illustrating the applied procedure for determining the residual stresses in the as deposited diamond coating structure is shown in Figure 5. The dimensions of the formed bulge after inclined impact test at certain conditions (diameter and height) are used as input parameters in the FEM-DRS algorithm for calculating the overall compressive residual stress in the diamond film structure (see Figure 5a). The overall principal stresses in σ_1 and σ_2 directions are equal and amount to approximately 10.2 GPa (see Figure 5b). Moreover, the corresponding compressive thermal residual stresses of the diamond coating in the same principal directions are calculated using the FEM-TRS algorithm (see Figure 5c). By subtracting the thermal residual stresses from the overall residual stresses, the remaining residual stresses in the diamond film structure, characterized as structural, were calculated (see Figure 5d) [7]. These stresses are generated during the film deposition and can be varied after the annealing process.

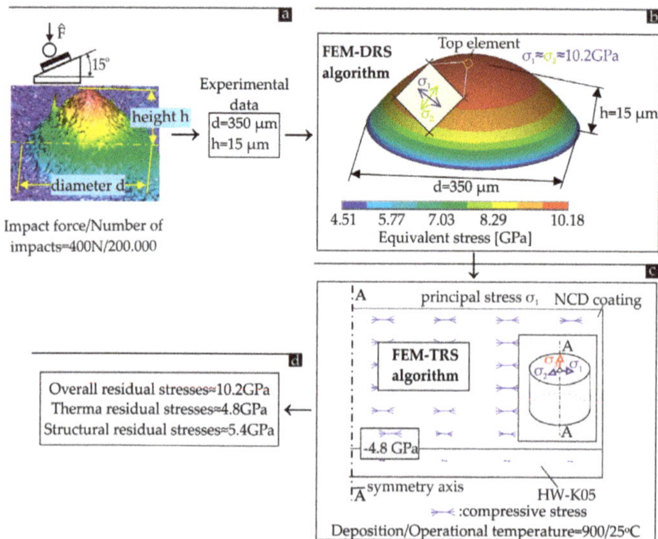

Figure 5. (**a**) Determination of the input parameters (diameter and height of the formed bugle) in the FEM-DRS algorithm; (**b**) calculation of the overall compressive stresses in the diamond film structure via FEM-DRS algorithm; (**c**) calculation of the thermal compressive stresses in the diamond film structure via FEM-TRS algorithm; (**d**) determination of the structural stresses in the diamond film structure.

Characteristic impact imprints after various number of impacts and loads at ambient temperature of the untreated as well as of the annealed diamond coated tools are shown in Figure 6. Taking into account the mechanical properties and the dimensions (diameter and height) of the formed bulges in each case as an input parameter in the FEM-model shown in Figure 3b (FEM-DRS algorithm), the overall residual stresses were calculated. According to the attained results, an increase of the annealing duration results in a significant reduction of the structural residual stresses. More specifically, after an annealing duration of 10 h, the structural residual are almost negligible and only thermal stresses exist in the film structure.

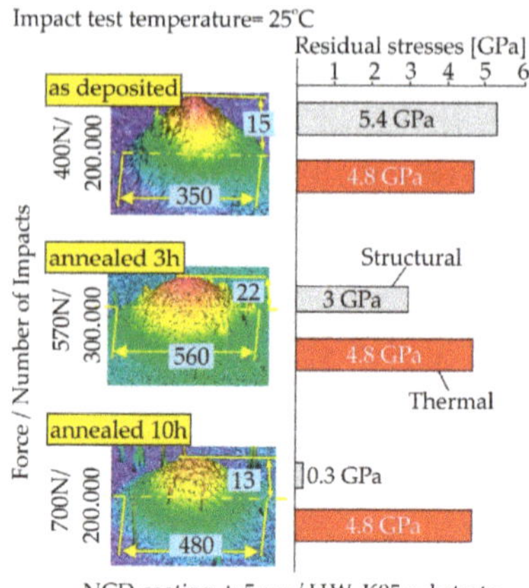

Figure 6. FEM-calculated thermal and structural stresses in the investigated coating cases based on the dimensions of the developed bulges after inclined impact tests and appropriate evaluation by using FEM-DRS and FEM-TRS algorithms.

3.3. Evaluation of the Interfacial Fatigue Strength of Diamond Coated Tools before and after Their Annealings

Inclined impact tests were conducted on the as deposited and annealed coated tools at various loads and number of impacts for evaluating their interfacial fatigue strength. During the conduct of the inclined impact test, shear stresses are developed in the diamond film-substrate interface. If the developed interface stresses lead to loads in the film substrate interface, which exceed the interfacial toughness of the film, an interface fatigue failure occurs. As a consequence, the high compressive residual stresses in the NCD structure are released hiking up the detached area at certain maximum height. The interfacial toughness of the diamond coatings depends strongly on the level of residual stresses in the film structure.

In Figure 7, characteristic topomorphies of the developed imprints on the untreated NCD coated insert at an impact load of 350 N after 100,000 as well as 300,000 and 400,000 impacts at ambient temperature are shown. After 100,000, there is no any sign of damage in the coating surface. An interface failure takes place after 300,000 impacts resulting to the bulge formation with a height of about 15 μm. The bugle is totally damaged and removed after only a further 100,000 impacts. Related topomorphies of impact imprints on the annealed coated inserts after 3 h and 10 h are illustrated at a load of 600 N and 650 N in Figures 8 and 9, respectively. The developed mechanism for diamond coating damage under the application of repetitive dynamic loads is similar including the coating detachment and its removal.

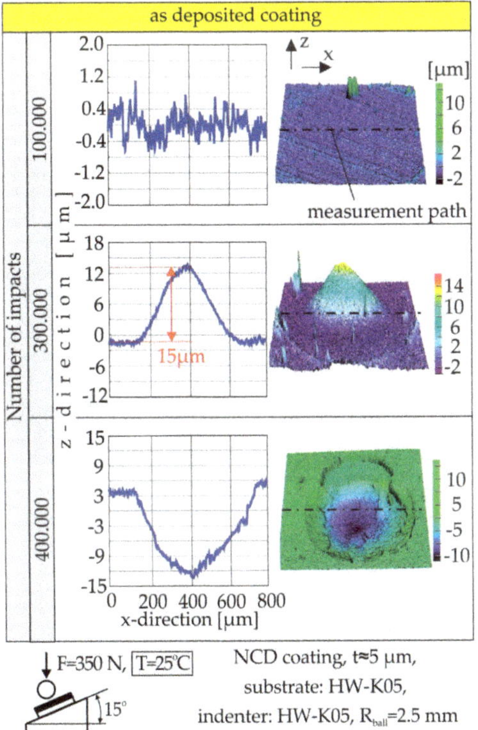

Figure 7. Impact test results on the as deposited nano-crystalline diamond (NCD) coated inserts at an impact load of 350 N after various number of impacts.

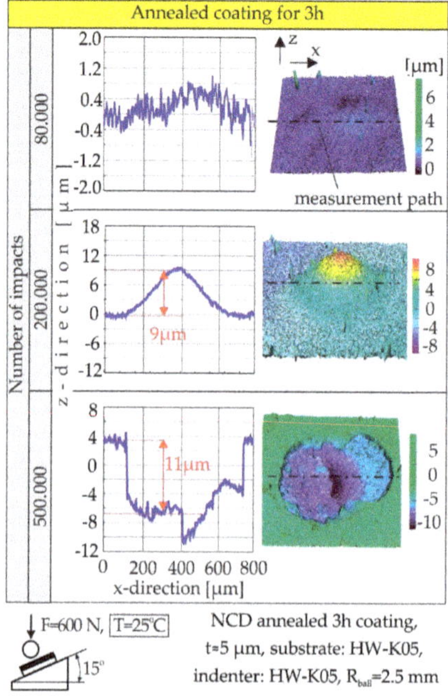

Figure 8. Impact test results on the annealed NCD coated inserts for 3 h at an impact load of 600 N after various number of impacts.

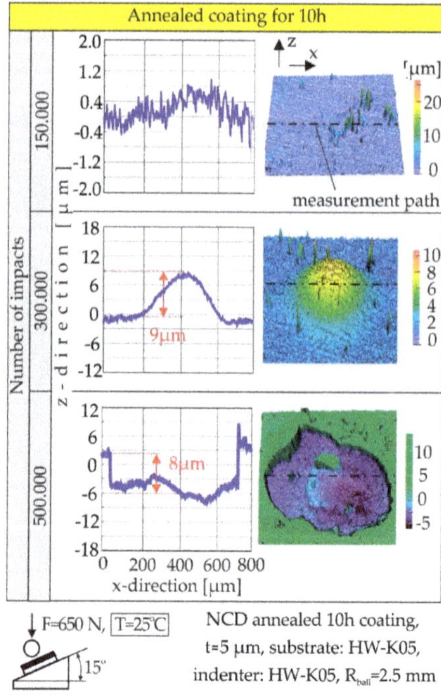

Figure 9. Impact test results on the annealed NCD coated inserts for 10 h at an impact load of 650 N after various number of impacts.

An overview concerning the occurring topomorphy of the as-deposited and annealed diamond coated tools after the conduct of the inclined impact tests at various loads and the number of impacts is shown in Figure 10a. All the impact tests were conducted at ambient temperature. After a certain combination of the impact load and number of impacts, bulges are formed. The experimentally detected maximum forces for avoiding the fatigue failure of the diamond coating-substrate interface after 10^6 impacts for the untreated as well as the annealed diamond coated tools after 3 h and 10 h amount to approximately 300 N, 525 N and 625 N (see Figure 10b). An impressive increase of threshold load for interfacial fatigue damage initiation after 10^6 impacts occurs when the structural stresses in the film structure are nullified due to the applied annealing process for 10 h.

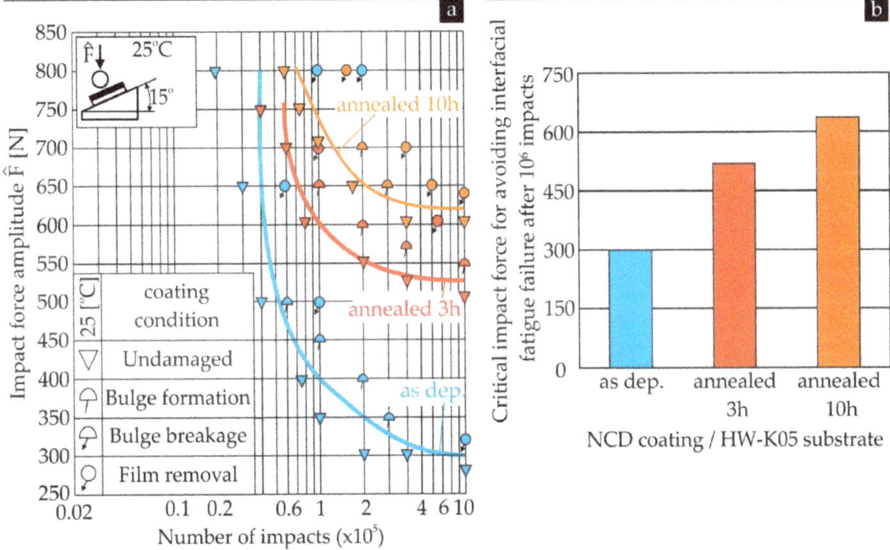

Figure 10. (**a**) NCD coating status after inclined impact tests at different combinations of impact load and number of impacts; (**b**) critical impact forces for avoiding interfacial damage after 10^6 impacts in the examined coating cases.

3.4. Explanation of the Attained Impact Test Results

For explaining the effect of the residual stresses of the NCD coating on its effective interfacial fatigue strength, the FEM-DRS algorithm was used (see Section 2). More specifically, two characteristic cases of NCD coatings were considered possessing overall residual stresses equal to approximately 10.2 GPa (as deposited case) and 5.1 GPa (annealed for 10 h) (see Figure 11a). In order to achieve a bulge height of 1 μm in both examined cases, the detached film region due to the interfacial fatigue failure is different. In the case of an NCD coating with residual stresses equal to approximately 10.2 GPa, the detached film area is associated with a diameter of only 60 μm, as shown in Figure 11b. However, a decrease of the residual stresses in the diamond film structure results in a significant augmentation of the detached film area for attaining the same coating lifting (see Figure 11b). Based on these results, it can be stated that the high level of residual stresses in NCD diamond coating triggers a coating lifting and removal when a fatigue damage occurs even in a restricted region in the coating-substrate interface, thus worsening the coating fatigue strength.

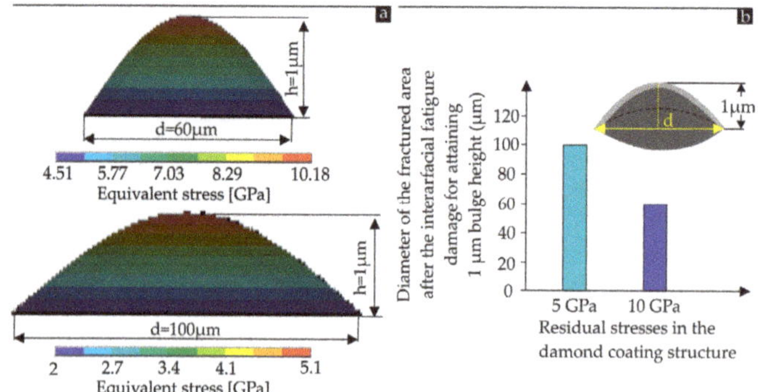

Figure 11. (**a**) Characteristic diamond coating cases for investigating the effect of the amount of their residual stresses on the occurring dimensions of the film bulges; (**b**) the occurring diameter of the fractured area of the investigated coating cases in order to attain 1 μm bulge height.

3.5. Wear Behavior of the Untreated and Annealed Diamond Coated Tools in Milling

The cutting performance of the untreated and annealed diamond coated tools was investigated in dry milling. In the way, the evolution of the wear was accelerated. Moreover, due to the structure of the aluminum foam used as workpiece material and the existence of especially hard phases, as already described in Figure 2c, intense dynamic load are developed in the coated cutting edge. In this way, a key-issue for attaining a sufficient tool life in milling is the interfacial fatigue strength of the diamond coated tool. After a prescribed number of successive cuts, the cutting insert wear was recorded. According to the cutting results shown in Figure 12a, in the case of the untreated coated tools, a flank wear width of 0.15 mm developed after only approximately 15,500 cuts. Moreover, the application of inserts annealed for 3 h at 400 °C enhances the coated tool life up to roughly 50,000 cuts. Coated inserts annealed for 10 h exhibit the best cutting performance reaching an impressive tool life of almost 78,000 cuts. Characteristic micro-graphs of the worn cutting edges in the examined annealing cases are shown in Figure 12b.

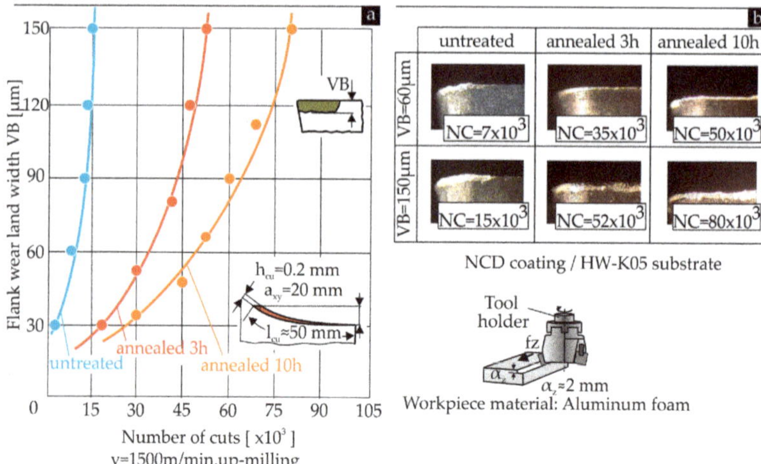

Figure 12. (**a**) Flank wear development versus the number of cuts of the investigated diamond coated inserts; (**b**) characteristic micro-graphs of the worn cutting edges in the examined coated inserts.

3.6. Review of the Obtained Results

A correlation among the diamond coating's residual stresses, the threshold impact load for interface fatigue endurance after 10^6 impacts as well as the tool life up to a flank wear width of 0.15 mm of the untreated (see Figure 13a) and annealed (see Figure 13b,c) coated tools at various conditions is conducted for having an overview of the achieved results. It is obvious that a decrease of the structural residual stresses in the diamond film structure due to annealing leads to a significant augmentation of the critical load for the initiation of diamond coating-substrate interface damage. As a result, the capability of the diamond coated cutting edge to withstand the dynamic loads in milling is enhanced resulting to an impressive enhancement of the coated tool life. It has to be pointed out that a percentile increase of approximately 400% of the coated tool life is attained when the structural residual stresses are totally removed from the initial diamond coating state after annealing.

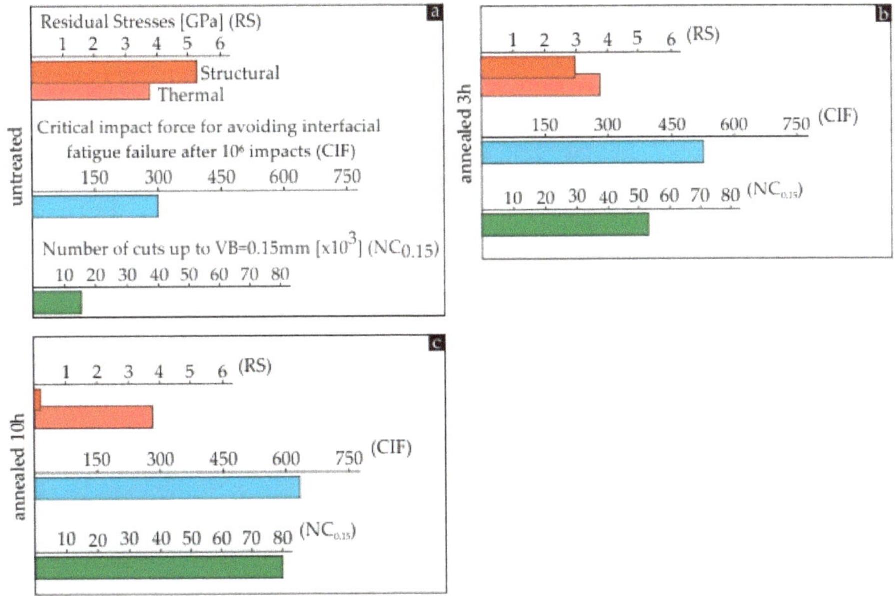

Figure 13. Correlation among NCD film residual stresses, interfacial fatigue strength as well as coated tool life in milling in the case of (**a**) an untreated coated tool; (**b**) an annealed one for 3 h at 400 °C; and (**c**) an annealed one for 10 h at 400 °C.

4. Conclusions

In the frame of this work, the potential to improve the wear behavior in the milling of NCD coated cemented carbide inserts via appropriate annealing was investigated. The obtained results show that an impressive enhancement of the effective interfacial fatigue strength and milling performance of diamond coated tools can be achieved by decreasing the structural residual stresses in the diamond film structure. More specifically, when the structural residual stresses are nullified via an appropriate adjustment of the annealing duration and temperature, the coated tool life is approximately four times larger than the initial one and thus the number of the worn tool replacements decreases significantly in machining processes. In this way, the conduct of the annealing process on NCD coated tools using appropriate conditions is highly recommended for improving their cutting performance.

Author Contributions: Conceptualization, G.S., F.S., T.K., and A.B.; methodology, G.S., F.S., T.K., and A.B.; software, G.S., F.S., T.K., and A.B.; investigation, G.S., F.S., T.K., and A.B.; data curation, T.K. and A.B.; writing—original draft preparation, G.S., F.S., T.K., and A.B.; writing—review and editing, G.S., F.S., T.K., and A.B.; supervision,

G.S. and F.S.; project administration, G.S. and F.S.; All authors have read and agreed to the published version of the manuscript.

Funding: This research is co-financed by Greece and the European Union (European Social Fund-ESF) through the Operational Programme "Human Resources Development, Education and Lifelong Learning 2014–2020" in the context of the project "Optimization of the cutting performance of nano- and micro-structured diamond coatings via appropriate thermal treatments" (MIS 5047906).

Conflicts of Interest: The authors declare no conflict of interest.

References

1. Haubner, R.; Kalss, W. Diamond deposition on hardmetal substrates—Comparison of substrate pre-treatments and industrial applications. *Int. J. Refract. Met. Hard Mater.* **2010**, *28*, 475–483. [CrossRef]
2. Ng, E.-G.; Szablewski, D.; Dumitrescu, M.; Elbestawi, M.A.; Sokolowski, J.H. High speed face milling of an aluminium silicon alloy casting. *CIRP Ann.* **2004**, *53*, 69–72. [CrossRef]
3. Uhlmann, E.; Reimers, W.; Byrne, F.; Klaus, M. Analysis of tool wear and residual stress of CVD diamond coated cemented carbide tools in the machining of aluminum silicon alloys. *J. Prod. Eng.* **2010**, *4*, 203–209. [CrossRef]
4. Polini, R. Chemically vapour deposited diamond coatings on cemented tungsten carbides: Substrate pretreatments, adhesion and cutting performance. *Thin Solid Films* **2016**, *515*, 4–13. [CrossRef]
5. Skordaris, G.; Bouzakis, K.-D.; Kotsanis, T.; Boumpakis, A.; Stergioudi, F.; Christofilos, D.; Lemmer, O.; Kölker, W.; Woda, M. Effect of the crystallinity of diamond coatings on cemented carbide inserts on their cutting performance in milling. *CIRP Ann.* **2019**, *68*, 65–68. [CrossRef]
6. Bouzakis, E.; Bouzakis, K.-D.; Skordaris, G.; Charalampous, P.; Kombogiannis, S.; Lemmer, O. Fatigue strength of diamond coating-substrate interface assessed by inclined impact tests at ambient and elevated temperatures. *Diam. Relat. Mater.* **2014**, *50*, 77–85. [CrossRef]
7. Skordaris, G.; Bouzakis, K.-D.; Charalampous, P.; Kotsanis, T.; Bouzakis, E.; Lemmer, O. Effect of structure and residual stresses of diamond coated cemented carbide tools on the film adhesion and developed wear mechanisms in milling. *CIRP Ann.* **2016**, *65*, 101–104. [CrossRef]
8. Bouzakis, K.-D.; Skordaris, G.; Bouzakis, E.; Charalampous, P.; Kotsanis, T.; Tasoulas, D.; Kombogiannis, S.; Lemmer, O. Effect of the interface fatigue strength of NCD coated hardmetal inserts on their cutting performance in milling. *Diam. Relat. Mater.* **2015**, *59*, 80–89. [CrossRef]
9. Polini, R.; Kumashiro, S.; Jackson, M.J.; Amar, M.; Ahmed, W.; Sein, H. A study of diamond synthesis by hot filament chemical vapor deposition on Nc coatings. *J. Mater. Eng. Perform.* **2005**, *15*, 218–222. [CrossRef]
10. Polini, R.; Barletta, M.; Cristofanilli, G. Wear resistance of nano- and micro-crystalline diamond coatings onto WC–Co with Cr/CrN interlayers. *Thin Solid Films* **2010**, *519*, 1629–1635. [CrossRef]
11. Silva, S.; Mammana, V.P.; Salvadori, M.C.; Monteiro, O.R.; Brown, L.G. WC-Co cutting tool inserts with diamond coatings. *Diam. Relat. Mater.* **1999**, *8*, 1913–1918. [CrossRef]
12. Slack, G.A.; Bartram, S.F. Thermal expansion of some diamondlike crystals. *J. Appl. Phys.* **1975**, *46*, 89–98.
13. Woehrl, N.; Hirte, T.; Posth, O.; Buck, V. Investigation of the coefficient of thermal expansion in nanocrystalline diamond films. *Diam. Relat. Mater.* **2009**, *18*, 224–228. [CrossRef]
14. Lee, D.G.; Fitz-Gerald, J.M.; Singh, R.K. Novel method for adherent diamond coatings on cemented carbide substrates. *Surf. Coat. Technol.* **1998**, *100–101*, 187–191. [CrossRef]
15. Staia, M.H.; D'Alessandria, M.; Quinto, D.T.; Roudet, F.; Marsal Astort, M. High-temperature tribological characterization of commercial TiAlN coatings. *J. Phys. Condens. Matter* **2006**, *18*, 1727–1736. [CrossRef] [PubMed]
16. Ferrari, A.C.; Robertson, J. Origin of the 1150-cm^{-1} Raman mode in nanocrystalline diamond. *Phys. Rev. B* **2001**, *63*, 121405. [CrossRef]
17. Bouzakis, K.-D.; Michailidis, N.; Hadjiyiannis, S.; Skordaris, G.; Erkens, G. The effect of specimen roughness and indenter tip geometry on the determination accuracy of thin hard coatings stress-strain laws by nanoindentation. *Mater. Charact.* **2002**, *49*, 149–156. [CrossRef]
18. Bouzakis, K.-D.; Skordaris, G.; Bouzakis, E.; Makrimallakis, S.; Kombogiannis, S.; Lemmer, O. Fatigue strength of diamond coatings' interface assessed by inclined impact test. *Surf. Coat. Technol.* **2013**, *237*, 135–141. [CrossRef]

19. Impact-BZ Ltd. Available online: www.impact-bz.com (accessed on 21 August 2020).
20. Teppernegg, T.; Klünsner, T.; Kremsner, C.; Tritremmel, C.; Czettl, C.; Puchegger, S.; Marsoner, S.; Pippan, R.; Ebner, R. High temperature mechanical properties of WC–Co hard metals. *Int. J. Refract. Met. Hard Mater.* **2016**, *56*, 139–144. [CrossRef]
21. Skordaris, G. Fatigue strength of diamond coating-substrate interface quantified by a dynamic simulation of the inclined impact test. *J. Mater. Eng. Perform.* **2014**, *23*, 3497–3504. [CrossRef]

MDPI
St. Alban-Anlage 66
4052 Basel
Switzerland
Tel. +41 61 683 77 34
Fax +41 61 302 89 18
www.mdpi.com

Coatings Editorial Office
E-mail: coatings@mdpi.com
www.mdpi.com/journal/coatings